I0055267

Handbook of Composite Materials and Structures

Handbook of Composite Materials and Structures

Edited by
Douglas Holliday

Larsen & Keller
www.larsen-keller.com

Handbook of Composite Materials and Structures
Edited by Douglas Holliday
ISBN: 978-1-63549-702-1 (Hardback)

© 2018 Larsen & Keller

⊟ Larsen & Keller

Published by Larsen and Keller Education,
5 Penn Plaza,
19th Floor,
New York, NY 10001, USA

Cataloging-in-Publication Data

Handbook of composite materials and structures / edited by Douglas Holliday.
    p. cm.
Includes bibliographical references and index.
ISBN 978-1-63549-702-1
    1. Composite materials. 2. Composite construction. 3. Materials. 4. Structural analysis (Engineering).
I. Holliday, Douglas.
TA418.9.C6 H36 2018
620.118--dc23

This book contains information obtained from authentic and highly regarded sources. All chapters are published with permission under the Creative Commons Attribution Share Alike License or equivalent. A wide variety of references are listed. Permissions and sources are indicated; for detailed attributions, please refer to the permissions page. Reasonable efforts have been made to publish reliable data and information, but the authors, editors and publisher cannot assume any responsibility for the vaility of all materials or the consequences of their use.

Trademark Notice: All trademarks used herein are the property of their respective owners. The use of any trademark in this text does not vest in the author or publisher any trademark ownership rights in such trademarks, nor does the use of such trademarks imply any affiliation with or endorsement of this book by such owners.

For more information regarding Larsen and Keller Education and its products, please visit the publisher's website www.larsen-keller.com

Table of Contents

Preface

The materials like mortars, metal composites, reinforced plastics, ceramic composites, concrete, etc. are the materials made by combining two different materials together to produce a third material which possesses properties and characteristics which are different from both its parent material. These new prepared materials are called composite materials. They are often more strong or light, or cheaper. This book is a valuable compilation of topics, ranging from the basic to the most complex theories and principles in the field of composite materials and their structure. The textbook is meant for students who are looking for an elaborate reference text on this field.

A detailed account of the significant topics covered in this book is provided below:

Chapter 1- Composite materials are the result of different materials with varying chemical and physical properties, when combined together, generate superior and unique materials. The common examples of composite materials are mortars, metal composites, ceramic composites, etc. This chapter is an overview of the subject matter incorporating all the major aspects of composite materials.

Chapter 2- Lamination is the method of manufacturing materials in layers. This process helps in improving the strength and stability of the material. Lamination is one of the methods that reduce stress and strain to a material, and this aspect is studied in the concept of strength of materials. The aspects elucidated in this chapter are of vital importance, and provide a better understanding of lamination.

Chapter 3- Solid mechanics focuses on the behaviour of solid and composite materials. It is very important in fields like civil, nuclear and mechanical engineering. Euler-Bernoulli Bearn equation is one of the application of solid mechanics. The section strategically encompasses and incorporates the major components and key concepts of solid mechanics, providing a complete understanding.

Chapter 4- Constitutive equations describe the relation between two physical quantities. It is used to derive the response of a material to external stimuli. Constitutive equations can either be phenomenological or can be derived from first principles. The topics discussed in the section are of great importance to broaden the existing knowledge on constitutive equations.

Chapter 5- Failure mechanics studies the situation where a system stops functioning properly. Some of the standards used to measure a damage or failure of a system are strength, stiffness, yielding, bending, resistance to lightening and resistance to hazardous environmental agents. This section will provide an integrated understanding of failure and damage mechanisms.

It gives me an immense pleasure to thank our entire team for their efforts. Finally in the end, I would like to thank my family and colleagues who have been a great source of inspiration and support.

Editor

A Brief Introduction to Composite Materials

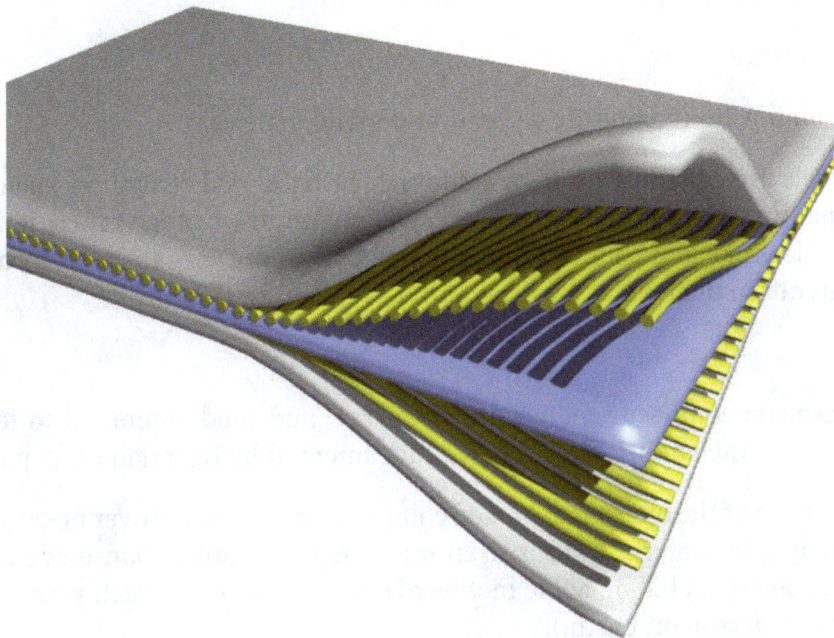

Composite materials are the result of different materials with varying chemical and physical properties, when combined together, generate superior and unique materials. The common examples of composite materials are mortars, metal composites, ceramic composites, etc. This chapter is an overview of the subject matter incorporating all the major aspects of composite materials.

Composite Material

Composites are formed by combining materials together to form an overall structure with properties that differ from the sum of the individual components

A Composite material (also called a composition material or shortened to composite, which is the common name) is a material made from two or more constituent materials with significantly different physical or chemical properties that, when combined, produce a material with characteristics different from the individual components. The individual components remain separate and distinct within the finished structure. The new material may be preferred for many reasons: common examples include materials which are stronger, lighter, or less expensive when compared to traditional materials. More recently, researchers have also begun to actively include sensing, actuation, computation and communication into composites, which are known as Robotic Materials.

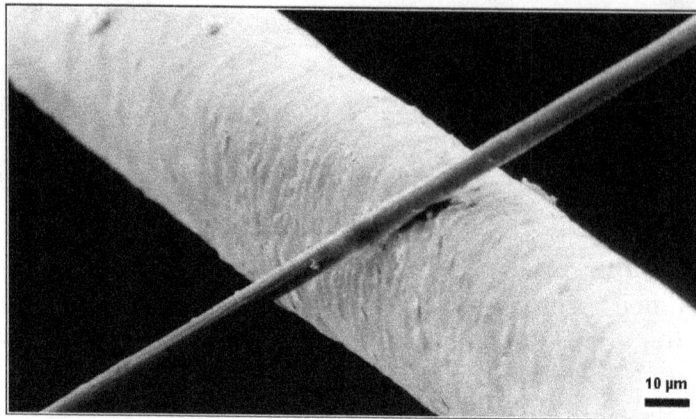

A black carbon fiber (used as a reinforcement component) compared to a human hair

Typical engineered composite materials include:

- mortars, concrete

- Reinforced plastics, such as fiber-reinforced polymer

- Metal composites

- Ceramic composites (composite ceramic and metal matrices)

Composite materials are generally used for buildings, bridges, and structures such as boat hulls, swimming pool panels, race car bodies, shower stalls, bathtubs, storage tanks, imitation granite and cultured marble sinks and countertops. The most advanced examples perform routinely on spacecraft and aircraft in demanding environments.

History

The earliest man-made composite materials were straw and mud combined to form bricks for building construction. Ancient brick-making was documented by Egyptian tomb paintings.

Wattle and daub is one of the oldest man-made composite materials, at over 6000 years old. Concrete is also a composite material, and is used more than any other man-made material in the world. As of 2006, about 7.5 billion cubic metres of concrete are made each year—more than one cubic metre for every person on Earth.

- Woody plants, both true wood from trees and such plants as palms and bamboo, yield natural composites that were used prehistorically by mankind and are still used widely in construction and scaffolding.

- Plywood 3400 BC by the Ancient Mesopotamians; gluing wood at different angles gives better properties than natural wood

- Cartonnage layers of linen or papyrus soaked in plaster dates to the First Intermediate Period of Egypt c. 2181–2055 BC and was used for death masks

- Cob (material) Mud Bricks, or Mud Walls, (using mud (clay) with straw or gravel as a binder) have been used for thousands of years.

- Concrete was described by Vitruvius, writing around 25 BC in his *Ten Books on Architecture*, distinguished types of aggregate appropriate for the preparation of lime mortars. For structural mortars, he recommended *pozzolana*, which were volcanic sands from the sandlike beds of Pozzuoli brownish-yellow-gray in colour near Naples and reddish-brown at Rome. Vitruvius specifies a ratio of 1 part lime to 3 parts pozzolana for cements used in buildings and a 1:2 ratio of lime to pulvis Puteolanus for underwater work, essentially the same ratio mixed today for concrete used at sea. Natural cement-stones, after burning, produced cements used in concretes from post-Roman times into the 20th century, with some properties superior to manufactured Portland cement.

- Papier-mâché, a composite of paper and glue, has been used for hundreds of years

- The first artificial fibre reinforced plastic was bakelite which dates to 1907, although natural polymers such as shellac predate it

- One of the most common and familiar composite is fiberglass, in which small glass fiber are embedded within a polymeric material (normally an epoxy or polyester). The glass fiber is relatively strong and stiff (but also brittle), whereas the polymer is ductile (but also weak and flexible). Thus the resulting fiberglass is relatively stiff, strong, flexible, and ductile.

Examples

Materials

Concrete is the most common artificial composite material of all and typically consists of loose stones (aggregate) held with a matrix of cement. Concrete is an inexpensive material, and will not compress or shatter even under quite a large compressive force. However, concrete cannot survive tensile loading (i.e., if stretched it will quickly break apart). Therefore, to give concrete the ability to resist being stretched, steel bars, which can resist high stretching forces, are often added to concrete to form reinforced concrete.

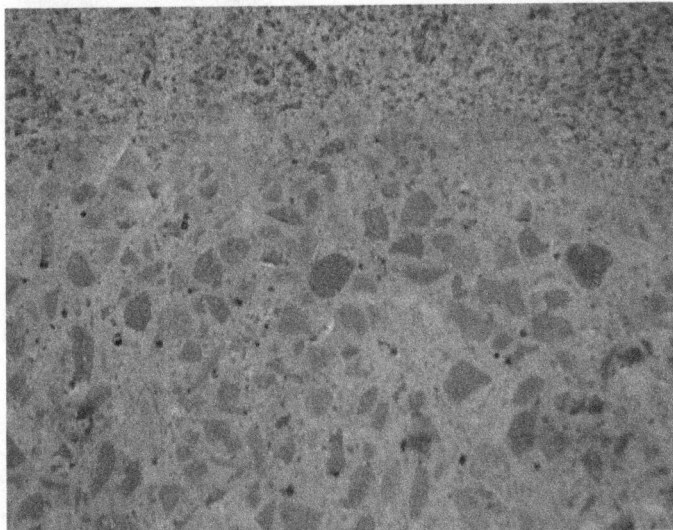

Concrete is a mixture of cement and aggregate, giving a robust, strong material that is very widely used.

Fibre-reinforced polymers or FRPs include carbon-fiber-reinforced polymer or CFRP, and glass-reinforced plastic or GRP. If classified by matrix then there are thermoplastic composites, short fiber thermoplastics, long fibre thermoplastics or long fibre-reinforced thermoplastics. There are numerous thermoset composites, including paper composite panels. Many advanced thermoset polymer matrix systems usually incorporate aramid fibre and carbon fibre in an epoxy resin matrix.

Plywood is used widely in construction

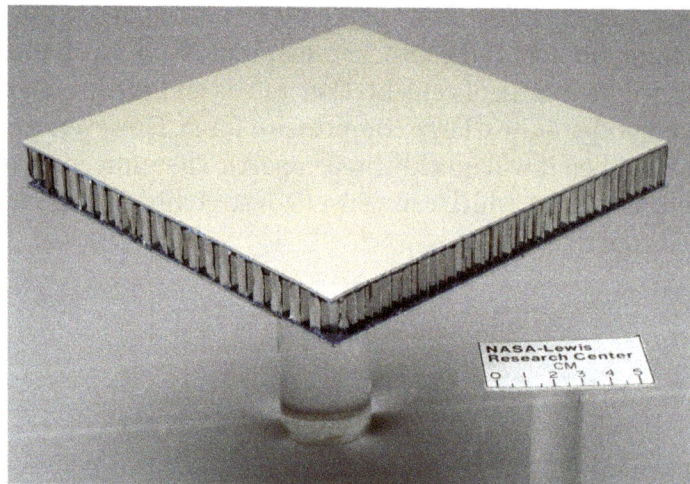

Composite sandwich structure panel used for testing at NASA

Shape memory polymer composites are high-performance composites, formulated using fibre or fabric reinforcement and shape memory polymer resin as the matrix. Since a shape memory polymer resin is used as the matrix, these composites have the ability to be easily manipulated into various configurations when they are heated above their activation temperatures and will exhibit high strength and stiffness at lower temperatures. They can also be reheated and reshaped repeatedly without losing their material properties. These composites are ideal for applications such as lightweight, rigid, deployable structures; rapid manufacturing; and dynamic reinforcement.

A. composites reinforced by particles;
B. composites reinforced by chopped strands;
C. unidirectional composites;
D. laminates;
E. fabric reinforced plastics;
F. honeycomb composite structure;

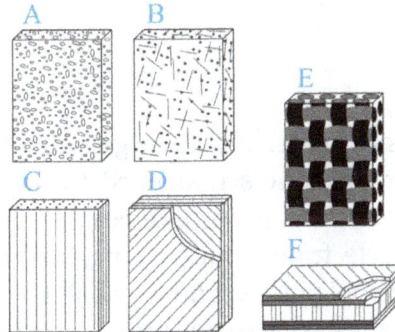

"Structural Integrity Analysis : Composites" (PDF).

High strain composites are another type of high-performance composites that are designed to perform in a high deformation setting and are often used in deployable systems where structural flexing is advantageous. Although high strain composites exhibit many similarities to shape memory polymers, their performance is generally dependent on the fiber layout as opposed to the resin content of the matrix.

Composites can also use metal fibres reinforcing other metals, as in metal matrix composites (MMC) or ceramic matrix composites (CMC), which includes bone (hydroxyapatite reinforced with collagen fibres), cermet (ceramic and metal) and concrete. Ceramic matrix composites are built primarily for fracture toughness, not for strength.

Organic matrix/ceramic aggregate composites include asphalt concrete, polymer concrete, mastic asphalt, mastic roller hybrid, dental composite, syntactic foam and mother of pearl. Chobham armour is a special type of composite armour used in military applications.

Additionally, thermoplastic composite materials can be formulated with specific metal powders resulting in materials with a density range from 2 g/cm^3 to 11 g/cm^3 (same density as lead). The most common name for this type of material is "high gravity compound" (HGC), although "lead replacement" is also used. These materials can be used in place of traditional materials such as aluminium, stainless steel, brass, bronze, copper, lead, and even tungsten in weighting, balancing (for example, modifying the centre of gravity of a tennis racquet), vibration damping, and radiation shielding applications. High density composites are an economically viable option when certain materials are deemed hazardous and are banned (such as lead) or when secondary operations costs (such as machining, finishing, or coating) are a factor.

A sandwich-structured composite is a special class of composite material that is fabricated by attaching two thin but stiff skins to a lightweight but thick core. The core material is normally low strength material, but its higher thickness provides the sandwich composite with high bending stiffness with overall low density.

Wood is a naturally occurring composite comprising cellulose fibres in a lignin and hemicellulose matrix. Engineered wood includes a wide variety of different products such as wood fibre board, plywood, oriented strand board, wood plastic composite (recycled wood fibre in polyethylene matrix), Pykrete (sawdust in ice matrix), Plastic-impregnated or laminated paper or textiles, Arborite, Formica (plastic) and Micarta. Other engineered laminate composites, such as Mallite, use a central core of end grain balsa wood, bonded to surface skins of light alloy or GRP. These generate low-weight, high rigidity materials.

Products

Fiber-reinforced composite materials have gained popularity (despite their generally high cost) in high-performance products that need to be lightweight, yet strong enough to take harsh loading conditions such as aerospace components (tails, wings, fuselages, propellers), boat and scull hulls, bicycle frames and racing car bodies. Other uses include fishing rods, storage tanks, swimming pool panels, and baseball bats. The new Boeing 787 structure including the wings and fuselage is composed largely of composites. Composite materials are also becoming more common in the realm of orthopedic surgery.And It is the most common hockey stick material.

Carbon composite is a key material in today's launch vehicles and heat shields for the re-entry phase of spacecraft. It is widely used in solar panel substrates, antenna reflectors and yokes of spacecraft. It is also used in payload adapters, inter-stage structures and heat shields of launch vehicles. Furthermore, disk brake systems of airplanes and racing cars are using carbon/carbon material, and the composite material with carbon fibers and silicon carbide matrix has been introduced in luxury vehicles and sports cars.

In 2006, a fiber-reinforced composite pool panel was introduced for in-ground swimming pools, residential as well as commercial, as a non-corrosive alternative to galvanized steel.

In 2007, an all-composite military Humvee was introduced by TPI Composites Inc and Armor Holdings Inc, the first all-composite military vehicle. By using composites the vehicle is lighter, allowing higher payloads. In 2008, carbon fiber and DuPont Kevlar (five times stronger than steel) were combined with enhanced thermoset resins to make military transit cases by ECS Composites creating 30-percent lighter cases with high strength.

Pipes and fittings for various purpose like transportation of potable water, fire-fighting, irrigation, seawater, desalinated water, chemical and industrial waste, and sewage are now manufactured in glass reinforced plastics.

Overview

Composites are made up of individual materials referred to as constituent materials. There are two main categories of constituent materials: matrix (binder) and reinforcement. At least one portion of each type is required. The matrix material surrounds and supports the reinforcement materials by maintaining their relative positions. The reinforcements impart their special mechanical and physical properties to enhance the matrix properties. A synergism produces material properties unavailable from the individual constituent materials, while the wide variety of matrix and strengthening materials allows the designer of the product or structure to choose an optimum combination.

Carbon fiber composite part.

Engineered composite materials must be formed to shape. The matrix material can be introduced to the reinforcement before or after the reinforcement material is placed into the mould cavity or onto the mould surface. The matrix material experiences a melding event, after which the part shape is essentially set. Depending upon the nature of the matrix material, this melding event can occur in various ways such as chemical polymerization for a thermoset polymer matrix, or solidification from the melted state for a thermoplastic polymer matrix composite.

A variety of moulding methods can be used according to the end-item design requirements. The principal factors impacting the methodology are the natures of the chosen matrix and reinforcement materials. Another important factor is the gross quantity of material to be produced. Large quantities can be used to justify high capital expenditures for rapid and automated manufacturing technology. Small production quantities are accommodated with lower capital expenditures but higher labour and tooling costs at a correspondingly slower rate.

Many commercially produced composites use a polymer matrix material often called a resin solution. There are many different polymers available depending upon the starting raw ingredients. There are several broad categories, each with numerous variations. The most common are known as polyester, vinyl ester, epoxy, phenolic, polyimide, polyamide, polypropylene, PEEK, and others. The reinforcement materials are often fibres but also commonly ground minerals. The various methods described below have been developed to reduce the resin content of the final product, or the fibre content is increased. As a rule of thumb, lay up results in a product containing 60% resin and 40% fibre, whereas vacuum infusion gives a final product with 40% resin and 60% fiber content. The strength of the product is greatly dependent on this ratio.

Martin Hubbe and Lucian A Lucia consider wood to be a natural composite of cellulose fibres in a matrix of lignin.

Constituents

Matrices

Organic

Polymers are common matrices (especially used for fiber reinforced plastics). Road surfaces are often made from asphalt concrete which uses bitumen as a matrix. Mud (wattle and daub) has seen extensive use. Typically, most common polymer-based composite materials, including fiberglass, carbon fiber, and Kevlar, include at least two parts, the substrate and the resin.

Polyester resin tends to have yellowish tint, and is suitable for most backyard projects. Its weaknesses are that it is UV sensitive and can tend to degrade over time, and thus generally is also coated to help preserve it. It is often used in the making of surfboards and for marine applications. Its hardener is a peroxide, often MEKP (methyl ethyl ketone peroxide). When the peroxide is mixed with the resin, it decomposes to generate free radicals, which initiate the curing reaction. Hardeners in these systems are commonly called catalysts, but since they do not re-appear unchanged at the end of the reaction, they do not fit the strictest chemical definition of a catalyst.

Vinylester resin tends to have a purplish to bluish to greenish tint. This resin has lower viscosity than polyester resin, and is more transparent. This resin is often billed as being fuel resistant, but will melt in contact with gasoline. This resin tends to be more resistant over time to degradation than polyester resin, and is more flexible. It uses the same hardeners as polyester resin (at a similar mix ratio) and the cost is approximately the same.

Epoxy resin is almost totally transparent when cured. In the aerospace industry, epoxy is used as a structural matrix material or as a structural glue.

Shape memory polymer (SMP) resins have varying visual characteristics depending on their formulation. These resins may be epoxy-based, which can be used for auto body and outdoor equipment repairs; cyanate-ester-based, which are used in space applications; and acrylate-based, which can be used in very cold temperature applications, such as for sensors that indicate whether perishable goods have warmed above a certain maximum temperature. These resins are unique in that their shape can be repeatedly changed by heating above their glass transition temperature (T_g). When heated, they become flexible and elastic, allowing for easy configuration. Once they are cooled, they will maintain their new shape. The resins will return to their original shapes when they are reheated above their T_g. The advantage of shape memory polymer resins is that they can be shaped and reshaped repeatedly without losing their material properties. These resins can be used in fabricating shape memory composites.

Traditional materials such as glues, muds have traditionally been used as matrices for papier-mâché and adobe.

Inorganic

Cement (concrete), metals, ceramics, and sometimes glasses are employed. Unusual matrices such as ice are sometime proposed as in pykecrete.

Reinforcements

Fiber

Differences in the way the fibers are laid out give different strengths and ease of manufacture

Reinforcement usually adds rigidity and greatly impedes crack propagation. Thin fibers can have very high strength, and provided they are mechanically well attached to the matrix they can greatly improve the composite's overall properties.

Fiber-reinforced composite materials can be divided into two main categories normally referred to as short fiber-reinforced materials and continuous fiber-reinforced materials. Continuous reinforced materials will often constitute a layered or laminated structure. The woven and continuous fiber styles are typically available in a variety of forms, being pre-impregnated with the given matrix (resin), dry, uni-directional tapes of various widths, plain weave, harness satins, braided, and stitched.

The short and long fibers are typically employed in compression moulding and sheet moulding operations. These come in the form of flakes, chips, and random mate (which can also be made from a continuous fibre laid in random fashion until the desired thickness of the ply / laminate is achieved).

Common fibers used for reinforcement include glass fibers, carbon fibers, cellulose (wood/paper fiber and straw) and high strength polymers for example aramid. Silicon carbide fibers are used for some high temperature applications.

Other Reinforcement

Concrete uses aggregate, and reinforced concrete additionally uses steel bars (rebar) to tension the concrete. Steel mesh or wires are also used in some glass and plastic products.

Cores

Many composite layup designs also include a co-curing or post-curing of the prepreg with various other media, such as honeycomb or foam. This is commonly called a sandwich structure. This is a more common layup for the manufacture of radomes, doors, cowlings, or non-structural parts.

Open- and closed-cell-structured foams like polyvinylchloride, polyurethane, polyethylene or polystyrene foams, balsa wood, syntactic foams, and honeycombs are commonly used core materials. Open- and closed-cell metal foam can also be used as core materials. Recently, 3D graphene structures (also called graphene foam) have also been employed as core structures. A recent review by Khurram and Xu et al., have provided the summary of the state-of-the-art techniques for fabrication of the 3D structure of graphene, and the examples of the use of these foam like structures as a core for their respective polymer composites.

Fabrication Methods

Fabrication of composite materials is accomplished by a wide variety of techniques, including:

- Advanced fiber placement (Automated fiber placement)
- Tailored fiber placement
- Fiberglass spray lay-up process
- Filament winding
- Lanxide process
- Tufting
- Z-pinning

Composite fabrication usually involves wetting, mixing or saturating the reinforcement with the matrix, and then causing the matrix to bind together (with heat or a chemical reaction) into a rigid structure. The operation is usually done in an open or closed forming mold, but the order and ways of introducing the ingredients varies considerably.

Mold Overview

Within a mold, the reinforcing and matrix materials are combined, compacted, and cured (processed) to undergo a melding event. After the melding event, the part shape is essentially set, although it can deform under certain process conditions. For a thermoset polymer matrix material, the melding event is a curing reaction that is initiated by the application of additional heat or chemical reactivity such as an organic peroxide. For a thermoplastic polymeric matrix material, the melding event is a solidification from the melted state. For a metal matrix material such as titanium foil, the melding event is a fusing at high pressure and a temperature near the melting point.

For many moulding methods, it is convenient to refer to one mould piece as a "lower" mould and another mould piece as an "upper" mould. Lower and upper refer to the different faces of the moulded panel, not the mould's configuration in space. In this convention, there is always a lower mould, and sometimes an upper mould. Part construction begins by applying materials to the lower mould. Lower mould and upper mould are more generalized descriptors than more common and specific terms such as male side, female side, a-side, b-side, tool side, bowl, hat, mandrel, etc. Continuous manufacturing uses a different nomenclature.

The moulded product is often referred to as a panel. For certain geometries and material combinations, it can be referred to as a casting. For certain continuous processes, it can be referred to as a profile.

Vacuum Bag Moulding

Vacuum bag moulding uses a flexible film to enclose the part and seal it from outside air. Vacuum bag material is available in a tube shape or a sheet of material. A vacuum is then drawn on the vacuum bag and atmospheric pressure compresses the part during the cure. When a tube shaped bag is used, the entire part can be enclosed within the bag. When using sheet bagging materials, the edges of the vacuum bag are sealed against the edges of the mould surface to enclose the part against an air-tight mould. When bagged in this way, the lower mold is a rigid structure and the upper surface of the part is formed by the flexible membrane vacuum bag. The flexible membrane can be a reusable silicone material or an extruded polymer film. After sealing the part inside the vacuum bag, a vacuum is drawn on the part (and held) during cure. This process can be performed at either ambient or elevated temperature with ambient atmospheric pressure acting upon the vacuum bag. A vacuum pump is typically used to draw a vacuum. An economical method of drawing a vacuum is with a venturi vacuum and air compressor.

A vacuum bag is a bag made of strong rubber-coated fabric or a polymer film used to compress the part during cure or hardening. In some applications the bag encloses the entire material, or in other applications a mold is used to form one face of the laminate with the bag being a single layer to seal to the outer edge of the mold face. When using a tube shaped bag, the ends of the bag are sealed and the air is drawn out of the bag through a nipple using a vacuum pump. As a result, uniform pressure approaching one atmosphere is applied to the surfaces of the object inside the bag, holding parts together while the adhesive cures. The entire bag may be placed in a temperature-controlled oven, oil bath or water bath and gently heated to accelerate curing.

Vacuum bagging is widely used in the composites industry as well. Carbon fiber fabric and fiberglass, along with resins and epoxies are common materials laminated together with a vacuum bag operation.

Woodworking Applications

In commercial woodworking facilities, vacuum bags are used to laminate curved and irregular shaped workpieces.

Typically, polyurethane or vinyl materials are used to make the bag. A tube shaped bag is open at both ends. The piece, or pieces to be glued are placed into the bag and the ends sealed. One method of sealing the open ends of the bag is by placing a clamp on each end of the bag. A plastic rod is laid across the end of the bag, the bag is then folded over the rod. A plastic sleeve with an opening in it, is then snapped over the rod. This procedure forms a seal at both ends of the bag, when the vacuum is ready to be drawn.

A "platen" is sometimes used inside the bag for the piece being glued to lie on. The platen has a series of small slots cut into it, to allow the air under it to be evacuated. The platen must have rounded edges and corners to prevent the vacuum from tearing the bag.

When a curved part is to be glued in a vacuum bag, it is important that the pieces being glued be placed over a solidly built form, or have an air bladder placed under the form. This air bladder has access to "free air" outside the bag. It is used to create an equal pressure under the form, preventing it from being crushed.

Pressure Bag Moulding

This process is related to vacuum bag molding in exactly the same way as it sounds. A solid female mold is used along with a flexible male mold. The reinforcement is placed inside the female mold with just enough resin to allow the fabric to stick in place (wet lay up). A measured amount of resin is then liberally brushed indiscriminately into the mold and the mold is then clamped to a machine that contains the male flexible mold. The flexible male membrane is then inflated with heated compressed air or possibly steam. The female mold can also be heated. Excess resin is forced out along with trapped air. This process is extensively used in the production of composite helmets due to the lower cost of unskilled labor. Cycle times for a helmet bag moulding machine vary from 20 to 45 minutes, but the finished shells require no further curing if the molds are heated.

Autoclave Moulding

A process using a two-sided mould set that forms both surfaces of the panel. On the lower side is a rigid mould and on the upper side is a flexible membrane made from silicone or an extruded polymer film such as nylon. Reinforcement materials can be placed manually or robotically. They include continuous fibre forms fashioned into textile constructions. Most often, they are pre-impregnated with the resin in the form of prepreg fabrics or unidirectional tapes. In some instances, a resin film is placed upon the lower mould and dry reinforcement is placed above. The upper mould is installed and vacuum is applied to the mould cavity. The assembly is placed into an autoclave. This process is generally performed at both elevated pressure and elevated temperature. The use of elevated pressure facilitates a high fibre volume fraction and low void content for maximum structural efficiency.

Resin Transfer Moulding (RTM)

RTM is a process using a rigid two-sided mould set that forms both surfaces of the panel. The mould is typically constructed from aluminum or steel, but composite molds are sometimes used. The two sides fit together to produce a mould cavity. The distinguishing feature of resin transfer moulding is that the reinforcement materials are placed into this cavity and the mould set is closed prior to the introduction of matrix material. Resin transfer moulding includes numerous varieties which differ in the mechanics of how the resin is introduced to the reinforcement in the mould cavity. These variations include everything from the RTM methods used in out of autoclave composite manufacturing for high-tech aerospace components to vacuum infusion to vacuum assisted resin transfer moulding (VARTM). This process can be performed at either ambient or elevated temperature.

Other Fabrication Methods

Other types of fabrication include press moulding, transfer moulding, pultrusion moulding, filament winding, casting, centrifugal casting, continuous casting and slip forming. There are also forming capabilities including CNC filament winding, vacuum infusion, wet lay-up, compression moulding, and thermoplastic moulding, to name a few. The use of curing ovens and paint booths is also needed for some projects.

Finishing Methods

The finishing of the composite parts is also critical in the final design. Many of these finishes will include rain-erosion coatings or polyurethane coatings.

Tooling

The mold and mold inserts are referred to as "tooling." The mold/tooling can be constructed from a variety of materials. Tooling materials include invar, steel, aluminium, reinforced silicone rubber, nickel, and carbon fiber. Selection of the tooling material is typically based on, but not limited to, the coefficient of thermal expansion, expected number of cycles, end item tolerance, desired or required surface condition, method of cure, glass transition temperature of the material being moulded, moulding method, matrix, cost and a variety of other considerations.

Physical Properties

The physical properties of composite materials are generally not isotropic (independent of direction of applied force) in nature, but rather are typically anisotropic (different depending on the direction of the applied force or load). For instance, the stiffness of a composite panel will often depend upon the orientation of the applied forces and/or moments. Panel stiffness is also dependent on the design of the panel. For instance, the fibre reinforcement and matrix used, the method of panel build, thermoset versus thermoplastic, type of weave, and orientation of fibre axis to the primary force.

In contrast, isotropic materials (for example, aluminium or steel), in standard wrought forms, typically have the same stiffness regardless of the directional orientation of the applied forces and/or moments.

The relationship between forces/moments and strains/curvatures for an isotropic material can be described with the following material properties: Young's Modulus, the shear Modulus and the Poisson's ratio, in relatively simple mathematical relationships. For the anisotropic material, it requires the mathematics of a second order tensor and up to 21 material property constants. For the special case of orthogonal isotropy, there are three different material property constants for each of Young's Modulus, Shear Modulus and Poisson's ratio—a total of 9 constants to describe the relationship between forces/moments and strains/curvatures.

Techniques that take advantage of the anisotropic properties of the materials include mortise and tenon joints (in natural composites such as wood) and Pi Joints in synthetic composites.

Failure

Shock, impact, or repeated cyclic stresses can cause the laminate to separate at the interface between two layers, a condition known as delamination. Individual fibres can separate from the matrix e.g. fibre pull-out.

Composites can fail on the microscopic or macroscopic scale. Compression failures can occur at both the macro scale or at each individual reinforcing fiber in compression buckling. Tension failures can be net section failures of the part or degradation of the composite at a microscopic scale where one or more of the layers in the composite fail in tension of the matrix or failure of the bond between the matrix and fibers.

Some composites are brittle and have little reserve strength beyond the initial onset of failure while others may have large deformations and have reserve energy absorbing capacity past the onset of damage. The variations in fibers and matrices that are available and the mixtures that can be made with blends leave a very broad range of properties that can be designed into a composite structure. The best known failure of a brittle ceramic matrix composite occurred when the carbon-carbon composite tile on the leading edge of the wing of the Space Shuttle Columbia fractured when impacted during take-off. It led to catastrophic break-up of the vehicle when it re-entered the Earth's atmosphere on 1 February 2003.

Compared to metals, composites have relatively poor bearing strength.

Testing

To aid in predicting and preventing failures, composites are tested before and after construction. Pre-construction testing may use finite element analysis (FEA) for ply-by-ply analysis of curved surfaces and predicting wrinkling, crimping and dimpling of composites. Materials may be tested during manufacturing and after construction through several nondestructive methods including ultrasonics, thermography, shearography and X-ray radiography, and laser bond inspection for NDT of relative bond strength integrity in a localized area.

Online Composites Portals

cdmHUB

cdmHUB is an online portal for composites resources, information, and networking. Launched in May 2013 at Purdue University, cdmHUB now hosts a rapidly growing collection of composites apps and commercial tools that run in the cloud and are accessible through a web browser. cdmHUB also provides a wide array of resources that help users learn, experience and interact with composites simulation tools and technology.

Composite materials have changed the world of materials revealing materials which are different from common heterogeneous materials. A composite material is a structural material that consists of two or more combined constituents which are combined at macroscopic level and are not soluble in each other. It should be understood that the aforesaid composite material is not the by-product of any chemical reaction between two or more of its constituents. One of its constituents is called the reinforcing phase and the other one, in which the reinforcing phase material is embedded, is called the matrix. The reinforcing phase material may be in the form of fibers, particles, or flakes (e.g. Glass fibers). The matrix phase materials are generally continuous (e.g. Epoxy resin). The matrix phase is light but weak. The reinforcing phase is strong and hard and may not be light in weight.

For example, in concrete reinforced with steel the matrix phase is concrete and the reinforcing phase is steel. In graphite/epoxy composites the graphite fibers are the reinforcing phase and the epoxy resin is the matrix phase.

A material shall be considered as a composite material if it satisfies the following conditions:

1. It is manufactured i.e., excluding naturally available composites.

2. It consists of two or more physically and/or chemically distinct, suitably arranged or distributed phases with an interface separating them.

3. It has characteristics that are not the replica of any of the components taken individually.

What Can Be Achieved by Forming a Composite Material

The following properties can be improved by forming a composite material:

❖ Strength (Stress at which a material fails)

❖ Stiffness (Resistance of a material to deformation)

❖ Wear & Corrosion resistance

❖ Fatigue life (long life due to repeated load)

❖ Thermal conductivity & Acoustical insulation

❖ Attractiveness and Weight reduction

What are the Roles of the Constituents of Composite Material

(i) Role of Reinforcements: Reinforcements give high strength, stiffness and other improved mechanical properties to the composites. Also their contribution to other properties such as the co-efficient of thermal expansion , conductivity etc is remarkable.

(ii) Role of Matrices: Even though having inferior properties than that of reinforcements, its physical presence is must;

❖ to give shape to the composite part

❖ to keep the fibers in place

❖ to transfer stresses to the fibers

❖ to protect the reinforcement from the environment, such as chemicals & moisture

❖ to protect the surface of the fibers from mechanical degradation

❖ to act as shielding from damage due to handling

Terminology

The following terms are frequently used in composite materials and hence it is necessary to know these terms.

- Staple fiber   : Represents discontinuous fiber

- Filament   : Represents a single continuous fiber

- Strand   : Represents a collection of untwisted fibers (filament) approximately 100 to 200 in numbers.

- Tow   : Represents bundle of untwisted filaments in large numbers, say 2000 to 12000 filaments

- Yarn   : Represents bundle of twisted fibers

- Sizes   : Represents a thin coating of chemical applied on filament surface to protect the fibers from damage and environmental effects (e.g., poly vinyle acetane)

- Coupling agents   : Used to get good bonding between fiber and matrix (e.g.,chrome complexes, silanes and titanes)

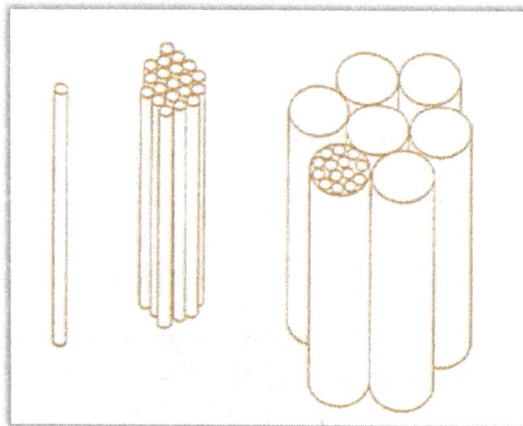

Fibers

Factors that Contribute to the Mechanical Performance of the Composites

As is mentioned earlier, the characteristics of the composite materials depend on the properties of both reinforcing phase as well as matrix phase. Therefore it is important to know the factors of the constituents of composite materials, which contribute to the performance of the composite materials.

(I) Factors that control the properties of fibers

(a) Length: The fibers can be long or short. Long, continuous fibers are easy to orient and process, but short fibers cannot be controlled fully for proper orientation. Long fibers provide many benefits over short fibers. These include high strength, impact resistance, low shrinkage, improved surface finish, and dimensional stability. However, short fibers provide low cost, easy to work with, and have fast cycle time fabrication procedures. Moreover using randomly oriented short fibers the isotropy behaviour may be achieved and uni directional composites exhibit non- isotropic material properties.

(b) Orientation: Fibers oriented in one direction give very high stiffness and strength in that direction. If the fibers are oriented in more than one direction, such as in a mat, there will be high stiffness and strength in the directions of the fiber orientations. Hence the fibers are usually oriented in directions where high stiffness and strength are required.

(c) Shape: Due to easiness in handling and manufacturing fibers, the most common shape of fibers is circular. But fibers are available in the form of square and rectangle also.

(d) Material: The material of the fiber directly influences the mechanical performance of a composite. Fibers are generally expected to have high elastic moduli and strength than the matrix materials. The fibers will also good functional properties like, high thermal resistance, fatigue resistance and impact resistance.

(ii) Matrix factors

Matrix materials have low mechanical properties compared to those of fibers. Yet the matrix influences many mechanical properties of the composite. These properties include

✓ Transverse modulus and strength

✓ Shear modulus and strength

✓ Compressive strength

✓ Inter-laminar shear strength

✓ Thermal expansion coefficient

✓ Thermal resistance and

✓ Fatigue strength

(i) Fiber-matrix interface

When the load is applied on a composite material, the load is directly carried by the matrix and it is transferred to the fibers from the matrix through fiber–matrix interface. So, it is clear that the load-transfer from the matrix to the fiber depends on the fiber-matrix interface. This interface may be formed by chemical, mechanical, and reaction bonding. In most cases, more than one type of bonding occurs.

(a) Chemical bonding: It is formed between the fiber surface and the matrix. Some fibers bond naturally to the matrix and others do not. Coupling agents are often added to form a chemical bond. Coupling agents are compounds applied to fiber surfaces to improve the bond between the fiber and the matrix.

(b) Mechanical bonding: Every material has some natural roughness on its surface. In composite materials, the roughness on the fiber surface causes interlocking between the fiber and the matrix leading to the formation a mechanical bond.

(c) Reaction bonding: It happens when molecules of the fiber and the matrix diffuse into each other only at the interface. Due to this inter-diffusion, a distinct interfacial layer, called the inter- phase, is created with different properties from that of the fiber or the matrix. Even though this thin interfacial layer helps to form a reaction bonding, it also develops microcracks in the fiber. These microcracks reduce the strength of the fiber and consequently that of the composite.

Fillers

In composite materials fillers are introduced for reducing the cost, for improving the physical or functional properties or to aid processing. Fillers are solid materials which are introduced on the matrix material for improving a specific property. Normally, fillers increase the modulus but reduces the strength and hence there must be always an optimal filler content. Fillers do not react with the matrix material, develop adequate bond with the matrix and do not absorb water or any other liquid. Normally, fillers are not used in most advanced composite structures, because fillers bring down the strength of the composite materials. Some of the fillers which are very commonly used in polyester resin and epoxy resins are given below.

Calcium carbonate, Silica powder, Talc, Clay are used in polyester resin to reduce the cost and for processing in SMC (Sheet molding compounds).

Sand and aggregates are used in polyester resin for making polymer concretes and marble chips are used to make artificial marbles. Titanium dioxide and carbon blacks are used in Polyester resin to give white and black colour respectively when used for gel coat.

Fused silica is used in epoxies to reduce coefficient of thermal expansion and mica is used to improve the thermal conductivity without affecting the electrical properties.

Aluminium trihydrate and Antimony trioxide are used in polyester and epoxies for improving fire retartant properties. Graphite is used to reduce the coefficient of friction and proving self lubricating property in these resins. Silicon carbide is used in these resins to increase the wear resistance by using them as a surface coat.

Additives

Additives are added to the polymer matrix for aiding the processing technique or altering some properties. They are added in small quantity (less than 5%) and the additives do not affect the mechanical properties due to their small quantity.

Hydroquinons is used as an inhibitor to inhibit the cure and prolong the shelf life. Parafin vax is used to prevent the evaporation of styrene from the coating surface. This act an air inhibitor. Tinorin, Benzophenos and Benzotriazoles are used as an UV stabilizer, to improve the resistance of UV rays. Aerosil powder is used to reduce the viscosity of the resin. Magnesium oxide, Calcium Oxide and Magnesium hydroxide are used to increase the viscosity of the resin. They act as a thickener in making SMC and BMC.

Pigments

Pigments are added to the resin to get composite products of different coloursIn wet lay up the pigment is added to the get coat and it is added to the moulding compound in compression moulding. The pigments readily mix in polyesters and in epoxies and phenolics these do not mix readily. There are organic and inorganic pigments. Inorganic pigments are fast and durable. The pigments are available in the form of pastes or powders. The paste form mixes faster than the powders.

Preprocessing of Composite Materials

FRP composites are prepared by the ingredients like, fibers, matrix, curing agents, fillers, pigments and additives, with different propositions. Some the ingredients are added in small quantity, which become cumbersome or time consuming while making large size products through wet moulding (Hand lay up) or through wet winding (filament winding process). To avoid this inconvenience, the raw materials are precomposed and brought in to an intermediate stage for further processing. There are several kind of precomposed materials and the details are given below.

Thermoset Moulding Compounds

The physical mixure of all the raw materials in the uncured resin system is called moulding compound and it is tack free. This can be a premix compound like DMC(Dough Moulding Compound) or a chemically thickened compound like SMC (Sheet Moulding Compound). These compounds are mostly used in compression moulding. Injection moulding grades are also available.

Pre impregnated Sheets or Prepregs: These are tack free rovings,tapes, clothes or mats of the reinforcement of fibers and reinforced in resin system which is semi cured, with suitable proportion. These prepregs are used in compression moulding or in filament winding.

Reinforced Thermoplastic Pellets

Short fibers or particulates are introduced in thermoplastic resins and pellets are prepared. The pellets are prepared by extruding the fibers and molten plastics by an extruder and chopped the extrudate in to short pellets. The pellets which are used for injection moulding will have the fiber length of 3 mm, which will pass through the nozzles of injection moulding or extruders easily.

There special kinds of injection moulding machines to use the long short fibers (length between 8 to 15 mm), which are preferred for achieving high strength.

Cowoven and Comingled Fiber Fabrics

The long reinforced and plastic are cowoven together to make thermoplastic prepregs of long length. During moulding process the plastic fibers melt and form the matrix. The disadvantage of this method is that when the adjacent fibers melt, there is larger gap between adjacent fibers and to avoid this fabrics made of comingled fiber bundles (have both the fibers) are used.

Thermoplastic Matrix Prepregs

The composite sheets or tapes are made by reinforcing the fibers in a thermoplastic matrix. The reinforcements can be in the form of continuous woven or non woven fiber mats or short fibers. The thermoplastic sheets are made by melt impregnation, slurry deposition, solution impregnation and by film slacking.

In melt impregnation the thermoplastics is melted and is impregnated in the fiber and then cooled.

In slurry deposition, the fiber is impregnated with the slurry of polymer and the liquid is evaporated out.

The solution impregnation is same as slurry deposition but the solution of the matrix is made use of.

In film slacking, the matrix is made in the form of a film, which is stacked between the reinforcements and melted to fuse in to the reinforcements.

Applications of Composite Materials

Nowadays there is no field of engineering without the foot-print of composite materials in some form. The following list is not the exhaustive list but to name a few. The application of composite materials can be broadly classified in to

- Aerospace applications
- Road and Rail transport applications
- Offshore accord water vehicles
- Building and other civil structures
- Chemical Industries
- Electrical, Electronics and communication applications
- Mechanical systems and machine elements
- Consumer durable products and sports applications
- Biomedical applications

High performance composites being costlier, is used mostly in defense applications where the performance is given high priority and not the cost.

Aircraft and Military Applications

The aerospace vehicles can be classified in to

1. Aircraft

2. Rockets and Missiles

3. Space crafts

Composites are mostly used in aerospace vehicles due to weight saving, high strength to weight ratio, high stiffness to weight ratio, and low fuel consumption. Fuel consumption is controlled by reducing the weight of the structure by using proper design criterion and by providing the aerodynamic shape to the structure. The most commonly used materials for weight reduction are some light weight metals and the alloys. Due to the increased speed, increase in usage and other specific requirements, new materials are essential and composite materials become the right choice. The primary advantage of composite materials is that these materials can be tailor made to meet any specific requirement and any complex shape can be easily made with composites. Few examples of aircraft and structural parts of aircraft which made use of composites is illustrated in the following pages.

Composites are used since weight reduction is very much critical for higher speeds.

| In 1969, Horizontal stabilizers of F-14 aircraft were made of boron fiber-reinforced epoxy skins. Carbon fiber-reinforced epoxy is used in wing, fuselage and empennage components. | The airframe of AV-8B and F-22 fighter aircraft contain nearly 25% by weight of carbon fiber- reinforced polymers. | The outer skin of B-2 and other stealth aircraft are almost made of carbon fiber-reinforced polymers. |

The list of parts of aircraft made of composite may be summarized as below:

- gliders
- Helicopter blades
- Transmission shafts
- Ailerons, rudders, elevators, flaps, spoilers etc
- Engine
- Cowlings
- Rocket
- Boosters

- Nozzles

- Antenna cover

- Fin and Fuselage portions

- Nose radome, doors, fairings

- Aircraft wing parts (skin, spars and stiffeners)

Automotive Applications

Based on the usage and shape the automobile vehicles, the automobile structure can be grouped in to three categories which include;

- Chassis components

- Body components

- Engine components

Composites are used in all the three major parts of the vehicles.

Composites are used in automobile industries due to the following reasons:

1) High specific strength (strength to weight ratio), due to which there is improvement in fuel efficiency.

2) Low wear and dear due to vibrations

3) Attractive styling

4) High resistance to damage

5) High resistance to corrosion, bio-degradation and extreme environmental conditions.

6) High thermal insulation and sound deadening properties

7) Easy repairing and maintenance free structural elements

E-glass fiber-reinforced sheet molding compound composites are used in the hood or door panels, radiator supports, bumper beams, roof frames, door frames, engine valve covers, timing chain covers, oil pans etc. Unileaf E-glass fiber-reinforced epoxy springs are used in place of multileaf steel springs.

In buses and trucks the GFRP composites are used to make, bus bodies, Roof panel, Engine Bonnet, Tail gates, mud wings, bumper bars, Window frames, doors, front grills and dash boards.

In cars the parts of car body, head lamp, rear window frame, tail lamp skirts, instrument panel, leaf spring and cabins are made of composites.

The body of refrigerated containers, petrol tankers, chemical tankers and milk tankers are made of composite panels.

In railways the parts of wash basin, battery box, seat and back rest, bathroom cubicles, window frames, guides, louvers, under carriage water tank and vacuum reservoirs are made of composite materials.

For all these applications mostly Glass reinforced composite products are used due to low cost and easy availability. Also the structural elements are mostly less stressed and hence glass fibers with polyester or Vinylester resins are used for these applications. The processing techniques are mostly Hand layup technique and compression moulding techniques for making these products.

Carbon fiber-reinforced polymers are used in the BMW M6 roof panel.

(a) BMW M6 car

(b) Race car made of composites

Carbon fiber-reinforced composite is extensively used in race cars. In Formula 1 race car, chassis, interior, and suspension components are made by the carbon FRC.

Like automotives, in railways also the following sections of train use composite materials.

- o Wagons

- o Fronts of power units

- o Doors, seats, interior panels

Marine Applications

Glass fiber composites are widely used in marine applications due to the following reasons:

1) Most of the conventionally used metals get corroded easily and timber and other natural materials decay quickly and glass fiber composites on the other hand has high corrosion resistant and can provide maintenance free service for longer duration.

2) GFRP boat hulls are maintenance free

3) Most of the complex shapes which provide buoyancy effect can be easily made.

4) GFRP boats can be made with less number of joints and structural elements and they are cost competitive.

5) Repair of composite products are easier

6) GFRP products are very light with high strength and stiffness and hence better substitute for steel materials.

Glass fiber-reinforced polyesters are used in sail boats, fishing boats, life boats, anti-marine ships, rescue crafts, hovercrafts and yachts. Due to higher strength of kevlar-49 fibers, these fibers are nowadays used for making boat hulls, decks, bulkheads, frames, masts and spars. Carbon fiber-reinforced epoxy is used in racing boats. Composites are used in naval ships, hulls, decks, bulkheads, masts, propulsion shafts, rudders, hunters, frigates, destroyers etc. Royal Swedish Navy is the largest composite ship producers in the world.

Most of the offshore structures which are made by composite materials include

- Building elements in the harbor
- Buoys of different types and
- Light houses

Sporting Goods

Due to light weight, high dampingproperties and design flexibility, FRC are widely used in making the following sports goods:

Tennis and Squash Rackets

- o Skis, Fishing rods
- o Hockey sticks
- o Arrows, Javelins
- o Baseball bats
- o Helmets

    o   Exercise equipments

    o   Athletic shoe soles and heels

Golf rackets are nowadays made of carbon fiber-reinforced epoxy due to its light weight. Glass fiber-reinforced epoxy is preferred over wood and aluminum in pole-vault poles because of its high strain energy storage capacity.

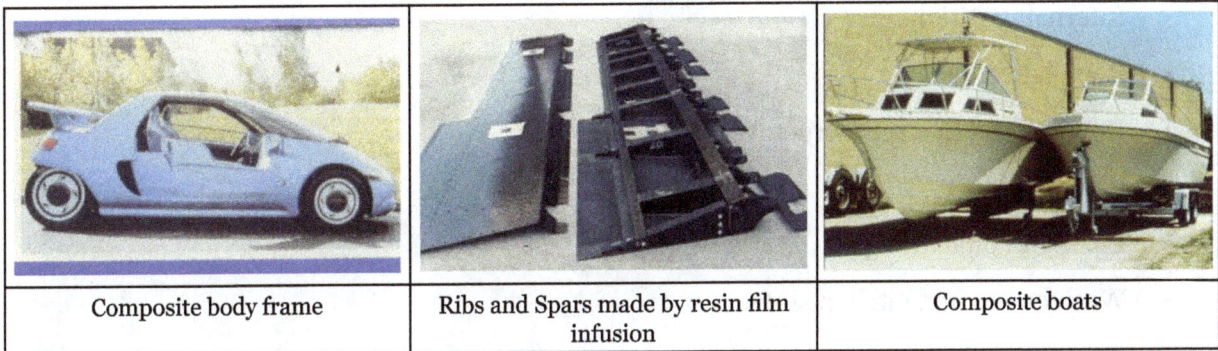

Composite body frame	Ribs and Spars made by resin film infusion	Composite boats

Infrastructure

Due to non-corrosion character, FRC is used in bridge construction. Due to light weight, FRC is preferred for constructing large bridges. Other areas in which composites are widely used, include

    o   Buildings

    o   Other civil structures

Building and Civil Engineering

- About 16% of GRP goes into this field
- Used for achieving complex architectural forms
- Costlier than many other construction materials
- A few kilogram of consumption of FRP can lead to fairly large market

Advantages in Civil Engineering Applications

- Used for portable prefabricated houses and offices
- Used for creating various architectural forms in buildings
- Used for replacing timber in buildings
- Translucent decorative and roofing panels
- Used for the renovation and remodelling of old buildings

Fibers like carbon, Kevlar and boron are not used in building constructions due to high cost and difficulty in getting these fibers. Some of FRP products in civil constructions are given below.

- Window frames
- Batgroom panels
- Cladding panels
- Internal partitions
- Roof light sheets
- Dooms and other roof structures
- Pipes and ducts
- Pipes and pipe fittings
- Wash basins and kitchen sinks
- Petrol station canopy
- Swimming pools
- Diving boards
- Translucent roofing sheets
- Doors and window frames
- Door panels
- Overhead water storage tank
- False ceiling panels
- Pipe lines for transporting oil and chemicals
- Lining of cannels

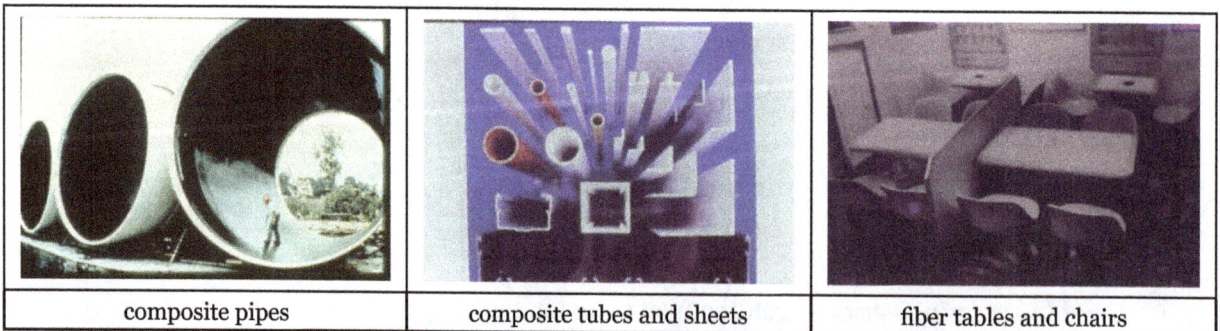

| composite pipes | composite tubes and sheets | fiber tables and chairs |

Advantages of Composites in Chemical Eqyipment and Corrosion Resistant Products

- Low cost of fabrication compared with other chemical resistant materials
- Used as anti-corrosive materials and also as construction materials.

- Thermosets are mainly used because of their adaptability to fabrication of large products.
- Thermoplastics without any reinforcements are used as lining materials.
- 50% to 2/3rd of GRP produced in India is used for the corrosive resistant product manufacture.

Applications

Tanks and Vessels

- Fuel storage tanks
- Reactors
- Boiling tubs

Towers and Columns

- Distillation column
- Cooling towers

Pumps, Fans, Blowers

- Centrifugal pumps
- Fans and blowers

Misc. Applications

- Valves and filters
- Water treatment equipment
- Electroplating equipment
- Photographic trays

Composites in Electrical, Electronics and Communication

- GRP are transparent to electromagnetic and sonar signals.
- GRP have properties like electrical insulation with high strength and light weight.
- Used for making printed circuit boards.
- GRP provides radio transparency and are easily mouldable to any shape.

Applications

Electrical Applications

- A.C. motor starter

- Transformer fuse block
- Cable ducts
- Switch activators
- Electrical separators
- Activator cases
- Low/high tension insulators

Electronic Applications

- Printed circuit boards
- Electromagnetic antennas
- Sonar and laser
- Radomes
- Radio and transistor housing

Radomes

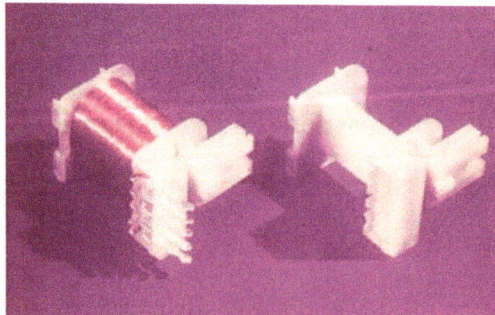

cable ducts

Machine Elements and Mechanical Engineering Applications

- Gears and bearings
- Various linkages of robots

- Fan housing

- Axial flow fan blades in power plants

- Cooling towers and cooling tower fan blades

- Wind mill blades

- Automobile leaf springs

- Automobile drive shafts

- Automobile engine

- Hydraulic cylinders

- Kinetic energy storing

- Springs and suspensions

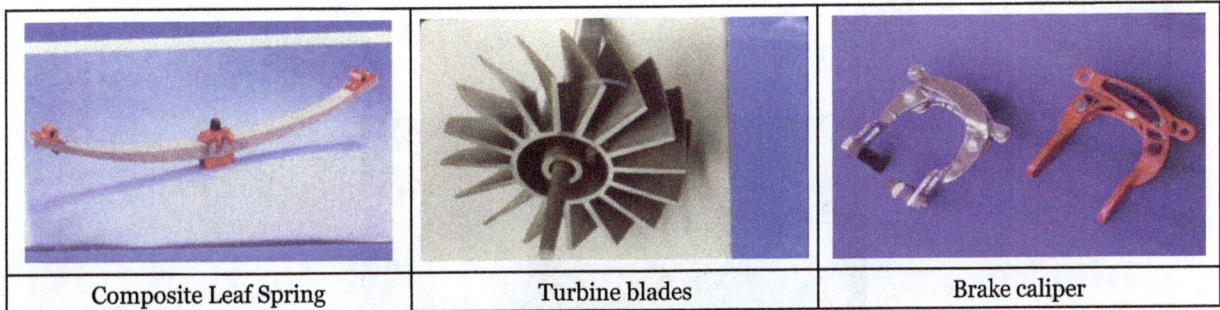

| Composite Leaf Spring | Turbine blades | Brake caliper |

Replacement of Various Human Body Parts with Composite Materials

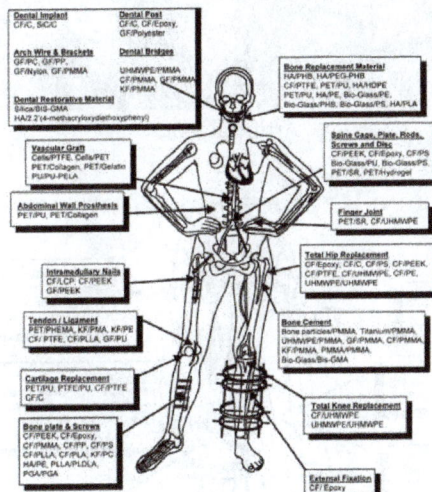

Composite material are promising in medicine

Implants are done by composites due to

- Mechanical properties similar to natural tissues

- Biological compatibility

- Can be formed into various shapes

- Bio stable material are used for long term implants PSU-HAP : Polysulphone-Hydroxyapatite

- PGLA-HAP : Poly (Lactide-co-glicol)-Hydroxyapatite

- Durability test is done by creep test

Rehabilitation Aid – Poly Propylene – carbon/glass Fiber

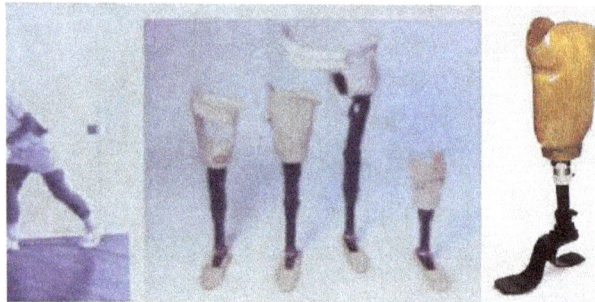

Rehabilitation aid

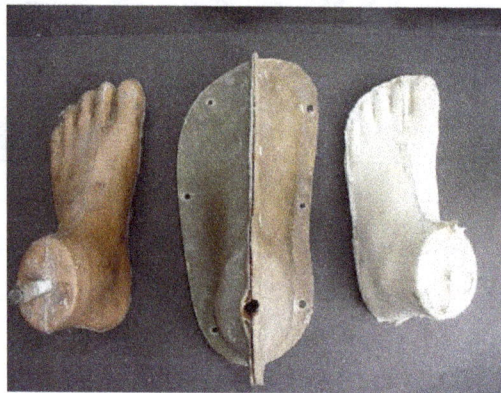

Artificial foot

Some of the Human Parts are Replaced by Composites

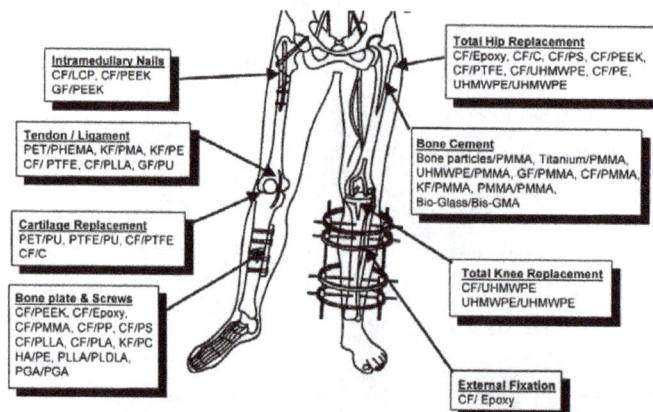

The parts of human leg being replaced by composites

Manufacturing Techniques

There are plenty of methods to cast a composite structure whether it is simple or complex, single or multiple. Each method has its own merits and limitations. Selection of particular manufacturing process is based on the type of matrix and fibers, temperature to form and cure the matrix, the geometry of the end product and cost effectiveness.

The two important parameters that control the manufacturing techniques are temperature and pressure. High temperature is required for the chemical reaction of resin to prevail whereas pressure is required for the highly viscous resin to flow into the fibers and to bind the fibers which are initially unbonded. The chemical reaction of resin forming cross linking is called curing. The time required to complete the curing is called the cure cycle.

Degree of Cure

The degree of cure at any time, t is defined by

$$\alpha_c = H/H_R$$

where, H - the amount of heat released in time t

H_R - heat of reaction

The degree of cure is determined experimentally using Differential Scanning Calorimeter (DSC). This detail will be useful in processing composites.

Gel Time

On curing, the viscosity of the matrix increases with increasing cure time and temperature. The rate of viscosity increase is low at the early stage of curing. After a threshold degree of cure is achieved, the resin viscosity increases at a very rapid rate. The time at which this occurs is called the gel time.

Gel Time Test

This test is conducted to determine the curing characteristics of a resin-catalyst combination. The procedure of the test is as below:

1. Take resin and catalyst and mix them thoroughly.

2. Pour the mix into a standard test tube which is suspended in a $82^{\circ}C$ water bath.

3. Insert a thermocouple in the test tube to monitor the temperature rise.

4. Record the time and the corresponding temperature rise.

5. Plot the graph Time vs Temperature.

6. From the graph, note down the point at which there is a sudden rise in temperature. The time corresponding to that point gives the gel time.

The typical gel time graph is shown in Figure.

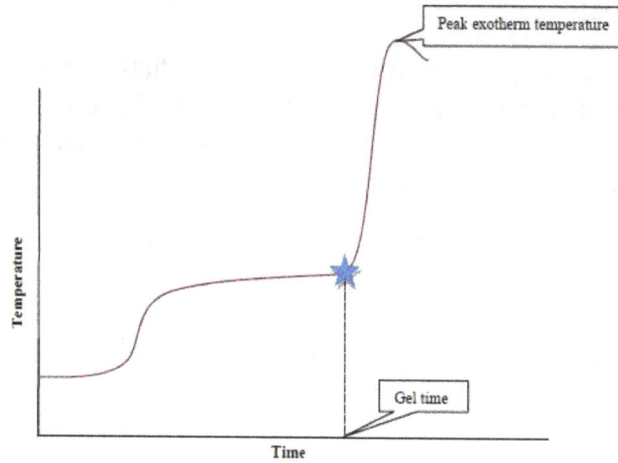

Temperature-time curve in Gel time test

Types of Composite Manufacturing

Composites are manufactured through different techniques. The techniques are chosen based on type of fiber, resin and the size of the product. Some of the commonly used manufacturing techniques are given below.

- Lay-up

  o   Hand lay-up

  o   Spray lay-up

  o   Prepreg Lay-Up

  o   Automatic tape lay-up

- Prepregs

- Compression molding

  o   Resin injection molding

  o   Incremental molding

  o   Stamp molding

  o   High-pressure compression molding

  o   Injection molding

- Bag molding

  o   Pressure bag molding

  o   Vacuum bag molding

- Autoclave molding

- Filament winding
    - Helical winding
    - Hoop winding
- Resin transfer molding (RTM)
    - Flexible RTM (FRTM)
    - Continuous RTM (CRTM)
    - Vacuum assisted RTM (VARTM)
    - High-speed RTM (HSRTM)
- Pultrusion
- Molding compounds
    - SMC (sheet molding compound)
    - BMC (bulk molding compound)
- Centrifugal Casting
- Extrusion method

(i) Prepregs:

Pre-impregnated fiber materials are called as prepregs. No thickening agent is used in making pre-pregs. Prepregs have a higher fiber content of 65%. They are available in both cloth or tape form. Usually, woven cloths are pre-impregnated, but woven rovings and chopped strand mats are also pre-impregnated.

Characteristics of Good Prepreg

1. The fiber to resin ratio should be high and should not vary from place to place.
2. Volatile contents and solvents should be minimum.
3. The prepreg should be flexible and tack free.
4. The material should have long storage life.
5. During moulding, the resin should be soften and flow filling the mold cavity should be without voids and defects.

Materials

Glass fiber is the most commonly used as reinforcing material, but other fibers like carbon fiber, boron have also been used. Epoxy and polyester resins are used as the impregnating agents.

Preparation of Prepregs

a) Method:

Prepregs can be made by basically two methods:

1. Wetting the glass fiber cloth with the resin and heating it to a B-stage of curing (partial curing) so that the material becomes tack free. After sometime, if the heat is withdrawn and the material is stored at -18°C the cross linking operation can be stopped. At the correct B-stage the cloth will be tack free and very flexible. The prepregs are slightly heated before processing to get soften and bond with the successive layers. After shaping (winding, press moulding), the material is heated to take it to the full cure.

2. If the matrix is in powder form it cannot go through the B-staging. In such cases, the resin is dissolved in a suitable solvent and brought to required viscosity.

b) Equipment:

The Machine used for manufacture of prepreg is called Tower. Fibers tensioned by tensioning device are passed through resin bath. It is then passed through a set of scrap bars to squeeze out the excess resin. The wetted fibers are then passed through drying oven, where temperature gradually increases. Volatiles are removed and resin reaches a tack free stage called B-stage. The prepreg fibers are covered by polythene sheet and completely rolled in aluminium foil.

c) Storage Condition:

The prepregs are stored in refrigerated chambers. The temperature of the storage area is important in improving the shelf life of the material. Moisture should be completely avoided. Shelf life is 6-8 months when stored at -18°C.

Evaluation of Prepregs

Parameters to be evaluated in cured state are:

1. Weight per unit area
2. Tackiness and durability
3. Resin content
4. Fiber content
5. Volatile content
6. Resin flow

Hand layup method:

It is the oldest molding method for making composite products. It requires no technical skill and no machinery. It is a low volume, labor intensive method suited especially for large components, such as boat hulls. A male and female half of the mould is commonly used in the hand lay-up process. A typical structure of hand lay-up product being made is shown in Figure.

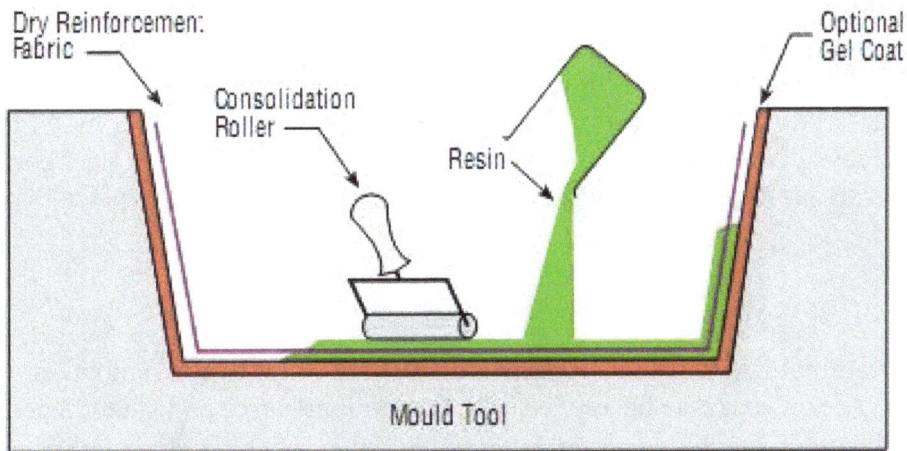

Hand layup method

Mould

The mould will have the shape of the product. In order to have a glossy or texture finish on the surface of the product, the mould surface also should have the respective finish. If the outer surface of the product to be smooth, the product is made inside a female mould. Likewise, if the inner side has to be smooth, the moulding is done over a male mould. The mould should be free from defects, since the imprint of any defect will be formed on the product.

Release Film or Layer

Since, the resins used are highly adhesive, the product may get stuck to the mould. So, a proper releasing mechanism should be incorporated. The release of the product can be affected.

> By the use of a release layer of wax or polyvinyl alcohol (PVA).

> By using a thin film like polyester film (Mylar).

Since, the Mylar sheet has to be fit into the mould profile, this method is not used for complex shapes.

Gel Coat

The gel coat gives the required finish of the product. It is usually a thin layer of resin about mm thickness applied on the outer surface of the product. The colour is obtained by adding appropriate pigments to the resin. The gel coat forms a protective layer that protects the glass fiber getting in contact with water and chemicals.

If the gel coat is too thin, the fiber pattern will become visible. If it is too thick, crazing and star crack can appear on the gel coat.

Surface Mat Layer

A surface mat layer will be placed beneath the gel coat layer. The fibers of the mat will not give high strength like reinforcement fibers, but the mat provides crack resistance and impact strength to the resin rich layer. It is an optional layer used only in specific cases.

Laminates of Glass Fiber

The glass fiber layer wetted with resin is laid up one after another to the required thickness and this finished material is called the laminate. The laminate gives the strength and rigidity to the product. Glass fiber in the chopped strand mat (CSM) is commonly used to get composite products. Woven roving, unidirectional and bi-directional mats are also used to get high strength composite products.

Finishing Surface Mat layer / Resin Coat

The glass fiber laminate provides a rough surface finish. In order to get a smoother surface, a surface mat layer or resin coat may be applied over the laminate layer and smoothened by placing a thin Mylar film layer.

Advantages

It is a low volume, labor intensive method suited for many products such as boat manufacturing, automotive components, ducts, tanks, furniture, corrosion resistant equipment etc.

No costly machinery is required.

Nearly all shapes and sizes can be made.

Colour and texture finish can be obtained by this hand lay-up method.

Limitations

The quality of the product depends on the skill of the operator.

It is not suitable for mass production of small products at high speeds.

It is difficult to get a void free composite product

Selection of Hand Lay-up As a Fabrication Process

The following conditions favor hand lay-up as the method of fabrication.

- Only one side need to have good smooth finish.
- The product is large in size and very complex in shape.
- Only a few numbers of mouldings are required.

Moulds

Open mould process of FRP fabrication makes use of either male or female mould. Open mould hand lay-up can be done in moulds made out of plaster of paris, wood, FRP, or metals.

Plaster of Paris mould is good for one or at most two pieces since the mould may break during the release of product. Wooden mould requires finishing work on every cycle of moulding. FRP moulds are ideal for complex shapes. When heating and pressing is required, metallic moulds are preferable.

Material Selection

Plaster of Paris, teak or rose wood, FRP, aluminium and die steel are good materials for making moulds.

Mould Thickness

Since, GRP is a costly material, the right thickness shall be chosen for GRP moulds. For a small complex shape product the mould thickness should be double the thickness of the product. For large size products in order to make economical, ribs or stiffeners are to be used rather than increasing the mould thickness. To avoid the warping of the mould, suitable flanges or stiffeners must be provided all round the edges.

Mould Trim Line Size

In hand lay-up the products are made with additional dimensions so that the product will have the required dimension after trimming. In case of cold pressing, extra space must be given to the mould for holding the excess resin squeezed out during pressing. This can be achieved by placing about 1/2" wide extra fiber mat all round which will act as a bleeder layer to absorb the resin.. The mould dimension is to be slightly larger than the product to hold the trim lines and bleeder layers. A bleeder layer is a synthetic material , available in variety of thicknesses and weights. It provides continuous air path for the pulling of vaccum from the composite products.

Mould Taper

For deep drawn products a taper has to be provided for ease release. It is a common practice that a 1 in 1000 taper for epoxy and a 1 in 100 taper for polyester are found to be adequate for easy release.

Split Mould Design

For large size and complex shaped products split moulds have to be provided. Flanges are provided at the two halves and it is connected by the bolted joints. The flange area should be 50% thicker than the mould shell thickness. A minimum flange width of 30mm with staggered bolting array may be provided.

Pattern and Pattern Making

For a small size product the mould is made by carving the wood. If the mould is made by GRP means it requires a pattern. For a large size and complex shape product like automobile body plaster of Paris is recommended.

Plaster of Paris Pattern and Moulds

The procedure for making mould and pattern making by plaster of Paris are the same. If the thickness is more than 15 to 25 mm, the plaster will take more time to set. So the construction should be such that there should not be any cracking. To prevent cracking chopped fibers will be added to the plaster.

Plaster moulds or patterns can be done by different methods which include, Rotational sweeping, linear sweeping, by using template and by sculpturing.

Rotational Sweeping

If the product has an axi-symmetric shape, the pattern can be shaped by rotating the template having the profile of the product to reproduce the shape.

Linear Sweeping

It is used for product having same profile along its length.

By Using Templates

If the product is of irregular shape, templates can be prepared to represent the product profile and to locate at their respective positions. The space between the templates is filled with plaster of Paris.

By Sculpturing

It is used to carve a profile from the cast plaster of Paris.

Since, the plaster of Paris has a porous surface it is smoothened with non-oil based putty like nitro cellulose putty or Duco putty. Finally, the Duco putty thinner is sprayed on the surface and polished with 400 grade emery sheet to have a smooth finish.

Mould Preparation and Application of Release Agents

The mould should be thoroughly cleaned and free from dirt's before the releasing agent is applied. Then, the mould surface is coated with silicone free wax (e.g. mansion polish). After some time the wax has to be removed to have a glassy finish on the mould surface. In certain cases release of the product is difficult with wax alone. So, a layer of poly vinyl alcohol (PVA) is applied. Since, PVA is water soluble material, 15% solution in water is applied with sponge. The brush application will leave the prints of brush lines so, sponge is preferable. After the water evaporates, a thin layer of PVA forms on the mould surface. The PVA layer must be completely dry before the gel coat is applied perhaps it will create wrinkles called 'elephant skin'. MEK or cellulose acetate, casein, carboxyl-methyl cellulose and methyl cellulose are the other film formers used as releasing agents.

Gel Coat Application

The gel coat resin is generally of the same as the matrix material used for making the composite product. The gel coat resin is prepared by adding 2 to 3 % of aerosol powder to the resin and stirring it well. The aerosol powder provides the required thixotropy. It is a property by which a liquid which remains as a thick viscous fluid with very little flow, but when some external force is applied (like stirring it well), it flows easily like a relatively low viscous fluid.

The viscosity of the resin can also be increased by adding fine calcium carbonate or other fillers. The filler percentage must be as low as possible otherwise, the cured resin will become brittle.

Pigments can be added with the resin to get the required colour. Titanium dioxide and carbon black will give white and black colour respectively. For products exposed to sunlight, UV stabilizers must be added to the gel coat resin. The required quantity of gel coat mix for a batch of same colours can be mixed at a time by adding pigments and accelerator to avoid colour variation.

Gel coat resin when applied must be free from air bubbles and dirt. About 600 gms. of resin will be required to give 0.5 mm thick gel coat on one sq. m. area. This will be applied as two coats. The second coat will be applied after the first coat is cured.

Surface Mat Layer

The surface mat layer must be applied only after the gel coat is cured. Otherwise the surface finish will be affected. The surface layer can be any one of the following.

> Glass fiber surface tissue mat.

> Polyester woven cloth of fine thickness

> Nylon woven cloth of fine thickness

A thin layer of resin is applied over the surface and the mat is wetted with brush. It may also be lightly rolled with roller to remove the air bubbles.

Lay-up of Laminates

The lay-up should start as soon as the gel coat layer is cured. The lamination should satisfy the following requirements:

a) The fiber layers should be uniformly placed and they should fit correctly into the contour of the product.

b) The fiber should not be damaged during lay-up

c) The fiber to resin ratio should be correctly maintained.

Preparation of the Resin Mix

The resin mix can be prepared at least one day ahead so that the entrapment of air bubbles escape before the lay-up begins. The mix consists of the resin, accelerator, fillers, and additives if any. The addition of accelerator to resin will not cause any cross linking until catalyst is added.

The mixing can be done by either manually using a paddle or by using an air operated mixer.

Vigorous stirring can cause entrapment of air bubbles therefore; mixing should be done at a very low rpm. The container in which resin mix is stored may be closed air tight to minimize the vaporization and loss of styrene.

Preparation of the Color Fiber Mat

The required number layers to obtain the thickness can be determined by taking into account the mat density and the glass-to-resin ratio by weight. The following points must be taken into account while preparing the mat:

Wherever joints are there, there should be a minimum overlap of 25 mm, in case of chopped strand and a 50 mm overlap is required in the case of woven roving mat.

Whenever, there is change in thickness the thickness must not abruptly change and instead it must gradually change.

Tools for Lay-up

○ Weighing balance - to weigh the chemicals.

○ Brushes - to apply resin for both gel coat application and for lamination.

○ Rollers - to remove the air bubbles and also for applying resin.

Long rollers are used to consolidate large areas but short rollers are used for corners and curved surfaces.

Mugs and small bowls - for taking the resin mix for lay-up.

Solvents

Solvents are required for cleaning the rollers and brushes during or after the lay-up sequence is over. Acetone or nitrocellulose thinner can be used as solvents.

Lamination Procedure

In the process of lamination a thin layer of resin is applied on the gel coat layer. Then, a chopped strand mat is placed over it. The resin is again applied over the mat by using brush to wet the mat. By using the roller the air bubbles are removed.

After the first layer is laid up, subsequent layers are laid in a similar manner. More than, 4 layers of resin and glass mat should not be applied without allowing the resin to cure at a time.

When WRM is laid up, CSM is used in between in order to increase the inter-laminar shear strength. The lay-up procedure for WRM and CSM are identical except that the resin used for WRM is half the quantity of that is needed for CSM.

Curing of Resin

The curing of resin process undergoes through four stages:

Gelation Stage

It is the stage at which the resin becomes tack free and unworkable. It depends on the percentage of catalyst and accelerator added. Normally, it takes 15 to 30 minutes to gel.

Green Stage

This is the stage at which the resin resembles to hard cheese which when pressed with the thumb it breaks up. The resin is considered to be set but not cured.

Cured Stage

It is the stage at which more than 90% of the cure is completed. The product can be released from the mould after this stage.

Fully Cured Stage

It is the stage at which the physical properties of the moulding are developed. Normally, it takes 5 to 10 days. At a fully-cured state, GRP will produce a metallic sound if it tapped with a coin.

Release of the Moulding from the Mould

This process should be done very carefully. Any of the following methods can be used for releasing the product.

Construct a 'grapple' point in the part so that when a hoist is engaged to lift upwards, the weight of the mould will cause to drop it off.

Wedges are inserted into the flange and by tapping it all around the mould; the two halves will be dropped.

Pultrusion

Pultruded composites consist of fibers predominantly in axial directions impregnated by resins in order form a most efficient composite product. Surface mats are used for surface appearance and also to improve chemical and weather resistance. Polyester resins are widely used in the pultruded products.

Pultrusion Process

The pultrusion process generally consists of pulling of roving/ mats through performing fixture to take its shape of the product and then heated where the section is cured continuously.

Description of Pultrusion Machine

The following are the facts of Pultrusion machine and the details of these facts are given below.

   a. Creel.

   b. Resin wet out tank

   c. Forming dies

   d. Heated matched metal die

   e. Puller or driving mechanism.

   f. Cut-off saw.

   g. Mandrel (for Hollow shapes)

Creel

Creel generally consists of bookcase type shelves where rovings from individual packages are pulled out for a resin bath. Metal book shelves are best since they can be grounded to avoid static charges produced. Vinyl tubes are installed to avoid the roving crossing over each other, as it generates "fuzz ball" to build up in the resin mix tank raising its viscosity.

Resin Wet Out Tank

The resin bath or wet out tank generally consists of sheet metal of aluminum through series of rolls. A grid or comb is attached at the entry and exits of resin wet out tank in order to maintain horizontal alignment and also to avoid the excessive resin.

Preforming Fixtures

These fixtures consolidate the reinforcements and move them closer to the final shape provided by the die. Generally, fluorocarbon or ultra high molecular weight polyurethanes are used as fixtures since these are easy to manufacture and also it is easy to clean it for later purpose.

Heated Dies

The chrome plated matched metal die maybe heated by electrical cartridges or by strip heaters. Thin sections are generally used by conduction of heat. In case of thick section the curing can be speeded up by using both radio frequency (RF) radiation and conductive heat.

Pulling Section

A pair of continuous caterpillar belts containing pads are used for pultrusions. A double set of cylinders with pad pullers can be synchronized for an intermittent pull.

Cut-Off Saw

A conventional saw with an abrasive or a continuous rim diamond wheel with coolant is generally used for cutting the desired product.

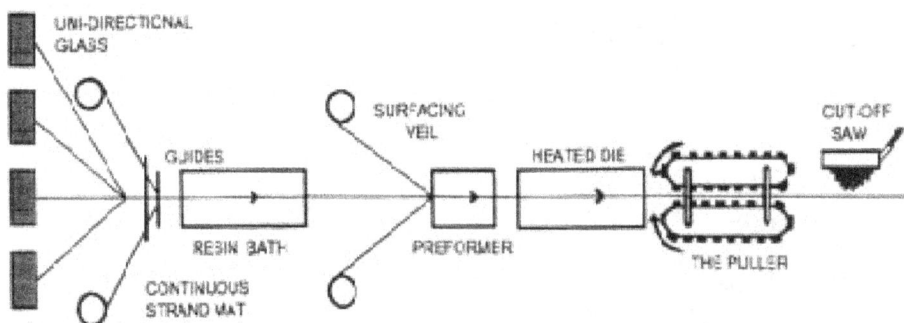

Pultrusion method

The pultruded sheet consists of both CS rovings as well as mat layers. Mat layers are added to increase the transverse strength. Generally, the matrix materials used in pultrusion are polyester and vinyl ester from thermoset polymers (epoxy has long cure time) and PEEK and polysulfone from thermoplastic polymers.

Pultruder sheet

The pultruded sheets are pulled through a liquid resin bath to thoroughly wet every fiber. The reinforcements are then guided and formed, or shaped, into the profile to be produced before entering a die. As the material progresses through the heated die, which is shaped to match the design profile, the resin changes from a liquid to a gel, and finally, into a cured, rigid plastic.

A pulling device grips the cured material and literally pulls the material through the die. It is the power source for the process. After the product passes through the puller, it is sawed into desired lengths. Although pultrusion is ideally suited for custom shapes, some standard products include solid rods, hollow tubes, flat sheets, hat sections bars, angles, channels, and I-beams.

Applications of Pultrusion

- Electrical application including transformers.

- Supports in bridges and structures.

- Automobiles.

- Pipes and rods.

Pultrusion Part Design

FRP property design criteria:

The design factor and load factor are to be considered before producing the product based on the application. It is necessary to depict its ultimate strength for safe operations. Apart from it various other properties like thermal and electrical properties are to be determined.

Table: Designs

Type of Loading	Minimum Design Factor
Static Short-term loads	2.0
Static Long-term loads	4.0
Variable or changing loads	4.0
Repeated loads, load reversal, fatigue loads	6.0
Impact loads	10.0

Pultrusion Part Design Principles

In this section the principles of product manufacturing and the handling factors of the product are discussed.

Table: Details of Pultrusion Process

Size	Shaping die and equipment pulling capacity influence size limitations
Shape	Straight, constant cross sections, some curved sections possible
Reinforcements	Fiberglass Carbon fiber Aramid fiber
Resin Systems	Polyester Vinyl ester Epoxy Silicones
Fiberglass Contents	Roving, 40-80% by weight Mat, 30-50% by weight Woven roving, 40-60% by weight
Mechanical Strengths	Medium to high, primarily unidirectional, approaching isotropic
Labor intensity	Low to medium
Mold cost	Low to medium
Production rate	Shape and thickness related

Structural Shapes

Early pultruded structural shapes were made to conform to standard steel practice. It was found out that since FRP shapes were heterogeneous materials and their shrinkage due to cure was subject to the type and quantity of resin used, warpage was a problem. These structural shapes should contain continuous strand mats as well as continuous rovings. An uneven number of plies of continuous strand mat are used with rovings placed between each two layers of mat.

Pultrusion Die Design

Pultrusion dies are considerably simpler in construction than most matched mould dies.

Die Steel

Any good tool steel can be used to make a Pultrusion die. Coated dies are in trend in order to withstand heat and also to avoid corrosion. Ceramic coated steel dies have been successfully used.

Bell Mouth Entrance

In order to assist the wet reinforcements to enter the mould a bell mouth is machined around the shape periphery. As the part size increases in width and area this bell mouth should be used for very large structural shapes.

Mounting Provisions

The die must be fastened to the heating platens with clamps or bolts. A less expensive and reliable method is to fasten the dies to the platens with bolts and angle clamps.

Die Surface Finish

All internal mould surfaces that see the FRP materials should have a good mould finish. Final polishing should be in longitudinal direction.

Chrome Plating

The internal areas of the Pultrusion die through which the materials are pulled must receive a hard chrome plate to provide a long working life for the die.

Heating

Pultrusion dies are to be heated with strip heaters, electrical cartridge heaters, or cored for hot oil. After it became useful to have several zones with different controlled temperatures and a different temperature at start up than during operation. The use of electrical cartridge heaters with thermocouple has now almost become a standard practice.

Cold Junction

A cold junction is used on the portion of the die that extends outside the heated platen area. Cooling water should enter the bottom cold junction port first and then the top plate as to insure that air pockets do not collect in the system.

Compression Molding

It is considered as the primary method of manufacturing for many structural automotive components, including road wheels, bumpers, and leaf springs. It is done by transforming sheet- molding compounds (SMC) into finished products in matched molds. It has the ability to produce parts of complex geometry in short periods of time. It allows the possibility of eliminating a number of secondary finishing operations, such as drilling, forming, and welding. Moreover the entire molding process can be automated.

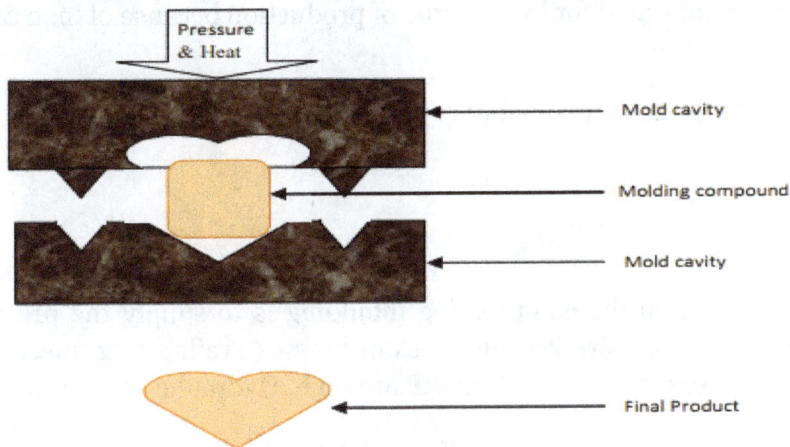

Compression molding

The molding compound is first placed in an open, heated mold cavity. The mold is then closed and pressure is applied to force the material to fill up the cavity. A hydraulic ram is often utilized to produce sufficient force during the molding process. Excess material is channeled away by the overflow grooves. The heat and pressure are maintained until the material is cured. The final part after the mold is removed. The molding pressure may vary from 1.4 to 34.5 MPa and the mold temperature is usually in the range of $130^{\circ}C$ to $160^{\circ}C$. To decrease the peak exotherm temperature which may cause burning and chemical degradation in the resin, filler may be added. The time to reach peak exotherm is also reduced with increasing filler content, thereby reducing the cure cycle. The cure time may also be reduced by preheat process.

There are two different types of compounds most frequently used in compression molding: Bulk Molding Compound (BMC) and Sheet Molding Compound (SMC). SMC costs higher but can be pre-cut to conform to the surface area of the mold.

Types of Compression Moulding

i) Hot pressing in which the moulding charge is heated while shaping.

ii) Cold pressing which uses a wet lay-up process and the product is pressed to the required shape, but cured without the application of heat.

Advantages

i) Good finish on both sides.

ii) Faster production.

iii) Uniform product quality.

iv) Less labor content.

v) Very little finishing operations required.

Disadvantages

i. This process is not suited for low volume of production because of high cost of moulds and press.

ii. The process is also not suitable for very large sized products.

Equipment

The Press

The function of the press in the compressing moulding is to supply the pressure required for moulding the products. There are various types of presses available eg. mechanical, pneumatic and hydraulic. Since, pressure required for GRP are high. Hydraulic presses are mostly used.

Moulds

Moulds give shapes to the moulding charge. Dimensional accuracy and surface finish of the moulded product depends mainly on the dimensional accuracy of the moulds and it has to be very high.

Mould surface should have high class surface finish and resistance to abrasion, since several thousand products have to be obtained from one mould.

Mould Materials

Requirements of mould materials are

   i   High strength

   ii High toughness

   iii High Hardness Value

   iv Good Abrasion Resistance

   v   Good Machinability

   vi Good weldability

   vii Good Polishability

Three general types of steel are used for mould construction

   i   Pre-toughened steel

   ii Case hardening steel

   iii Air hardening steel

Alloy steel AISI-4140 or its equivalent IS-40 C 1 Mo 28 or EN19C pre-hardened to Rockwell C30-32 is used for high class moulds.

Types of Moulds

There are five standard designs for compression mould cavities and forces, these are

   1. Flash type mould:

This design is not recommended, since parts produced may be of poor quality. However, it may be used for large parts made form BMC or SMC.

   2. Fully positive mould:

It is used for large deep draw where maximum density is required.

   3. Landed positive mould:

These are multi-cavity moulds. Multi-cavity mould may be this type of mould. Vents are incorporated on the force to permit maximum density.

   4. Semi-positive vertical flash mould:

These are more suited for automatic moulding

5. Semi-positive vertical flash mould:

This is used when no visual flash line mark is permitted on the moulded parts. Mold costs are more because of two areas of proper fit between force and cavity.

Resin Transfer Molding (RTM)

It is also called as liquid molding. It is a low pressure closed molding process for moderate volume production quantities. Dry continuous strand mats and woven reinforcements are laid up in the bottom half mold. Preformed glass reinforcements are often used for complex mold shapes. The mold is closed and clamped, and a low viscosity, catalyzed resin is pumped in, displacing the air through strategically located vents.The injection pressure of resin is in the range of 70-700 kPa.

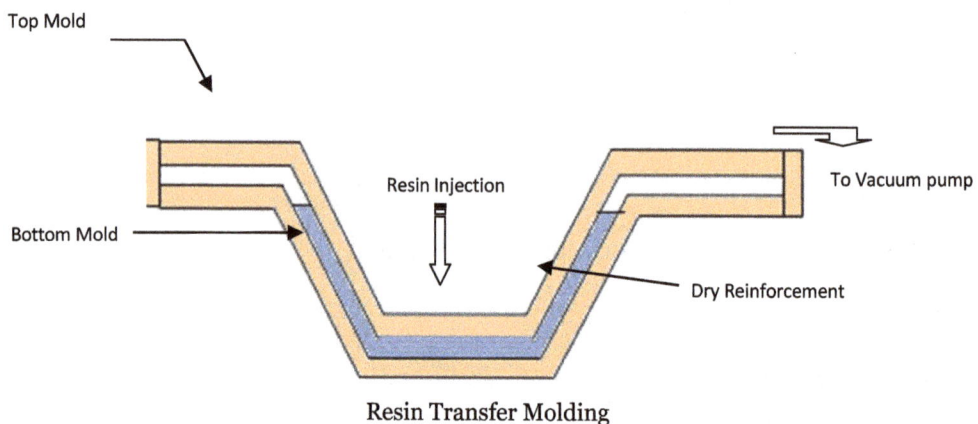

Resin Transfer Molding

Advantages and Limitations

Unlike in hand layup, RTM process gives better control on product thickness and good finish on both sides. It is not essential to have metallic moulds because the product curing is generally done under ambient temperature. By applying gel coats on both sides, the product will have a smoother finish on both sides.

When the injection pressure is increased, as in very closely packed fibers, there is a tendency for fiber wash. This tendency can be countered by using continuous strand mats or special woven performs. Inserting of wood, foam or metal will reduce the secondary bonding. Other advantages can be listed as follows.

1. Controlled usage of fiber and resin reduces the material wastage and unit cost

2. A variety of mould shapes and sizes can be moulded sequentially because of the mobile pumping unit.

3. Styrene emission is practically eliminated during resin transfer into the mould.

   There are few limitations which require special attention. Some of the limitations are given below:

1. Since, this process can develop pressures up to 5 to 10 bars, tool rigidity and clamping techniques have to be designed for such pressures.

2. Handling of large and heavy moulds requires adequate lifting equipments.

3. Unlike in compression moulding, post trimming is required for this process.

Process Equipment And Tools

Types of RTM Machines

The machines used for RTM include a mixing head attached to a nozzle, a pumping unit, and a solvent flushing unit. The pumping unit generates the pressure to inject the resin through the layers of reinforcement. The solvent flushing unit pumps solvent such as acetone to clean the mixing and injection chamber free of resin.

There are three types of RTM injection equipments based on position of mixing of catalyst with resin.

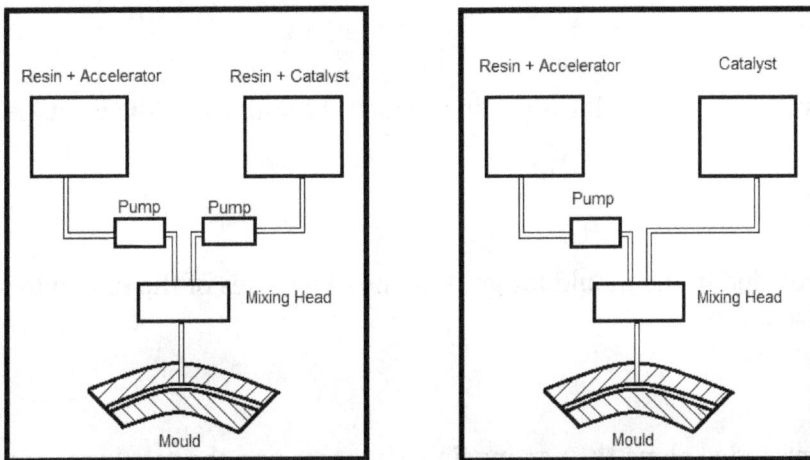

| A two pot RTM machine | A catalyst dispersing type RTM machine |
| (a) Two pot RTM machine | (b) catalyst dispersed RTM machine |

a. Two pot system

This system has two equal volume containers or pots. In one of these pots the resin is mixed with accelerator. In the other pot the resin is mixed with the catalyst. Two pumps are used to pump these mixtures to the injection points where they are mixed well in the mixing head.

b. Catalyst injection system

In this system the catalyst is not mixed with the resin until it reaches the entry pot attached to the mould. The resin mixed with accelerator is pumped into the injection chamber. The catalyst is taken separately into the chamber by means of controlling valve. The advantage of this system is that the gel and cure time can be controlled by varying the amount of catalyst added.

c. Pre-mixing system

This is a simple process by mixing the resin, accelerator, and catalyst in a vessel directly and injecting the mixture into the mould. A thick walled airtight metallic cylinder provided with inlet and outlet holes is taken. The injection is carried out through the outlet by means of compressed air. The cylinder has to be washed periodically with acetone to prevent clogging by cured resin.

Moulds and Mould Design

Mould Structure

The mould essentially consists of male and female halves, clamped by a clamping arrangement. Other parts include the injection ports, the air vents, the guide pins, and the gasket along the partition line.

Injection Ports

Injection port is the nozzle through which the resin is injected into the mould. The correct location of injection port is very much essential to ensure proper filling of the mould. As far as possible, the injection port must be located at the middle so that the resin flows radially to the periphery.

Air Vents

Air vents are provided at suitable locations in the mould for allowing the volatiles and trapped air from the part.

Guide Pins

Guide pins are provided in the mould for guiding the two halves of the mould to a perfect closure without lateral displacement.

Gaskets

Sealing gasket is provided along the parting line while crossing the mould for preventing the flow of resin through the parting line. Neoprene and silicone can be used as the gaskets.

Moulding Process

Mould Preparation

The two halves of the mould are cleaned and the dust must be removed from the surface. Wax polish is then applied which helps in easy release of the mould after curing. Over the layer of wax, a film of PVA is applied to aid the release. The disposable inlet and outlet port and air vents are then fitted in position.

Gel Coating

A layer of gel coat with appropriate pigment is applied on the surface of both halves of the mould. The gel coat thickness should not exceed 0.5mm.

Fiber Packing and Mould Closure

The calculated quantity of fiber is placed inside the mould. Wherever, the overlap comes, a 25 to 35 mm overlap must be given. The plies near the inlet port can be stitched together otherwise the fiber wash can occur due to injection pressure. The inserts should be placed correctly before the mould is closed. The clamping of mould has to be tight enough to withstand the injection pressure.

Resin Injection and Curing

The resin is then injected to the mould using an RTM machine at a calculated pressure. Care must be taken to see that the right quantity of catalyst is dosed into the resin stream and no gelling occurs during pumping. The mixture head has to be pumped with acetone at 15 minutes interval so that the resin does not set within the mixture head.

Demoulding and Cleaning

The mould is left undisturbed until the resin is fully cured. For products with large thickness, the high exotherm may lead to degradation of resin hence mould cooling is necessary to reduce the heat. Demoulding is done by removing the clamps and by releasing the mould without any damage to the mould. The product and the mould are then cleaned thoroughly. The product can be polished by using emery paper.

Mould Time Cycle

Total moulding time is given by the relation:

$$T_T = T_{mf} = T_{gel} = T_c = T_u = T_{cl} + T_p + T_{fp}$$

T_{mf} = mould fill up time

T_{gel} = gel time

T_c = cure time

T_u = un mould time

T_{cl} = cleaning time

T_p = preparation time

T_{fp} = fiber packing time

Filament Winding

Filament winding consists of winding resin impregnated fibers or rovings of glass, aramid, or carbon on a rotating mandrel in predetermined patterns.

The method makes void free product possible and gives high fiber volume ratio up to 80%. In the wet method, the fiber picks up the low viscosity resin either by passing through a trough or from a metered application system. In the dry method, the reinforcement is in the pre impregnated form.

After the layers are wound, the component is cured and removed from the mandrel. This method is used to produce pressure vessels, rocket motor cases, tanks, ducting, golf club shafts, and fishing rods and to manufacture prepregs. Thermoset resins used in filament wound parts include polyesters, vinyl esters, epoxies, and phenolics.

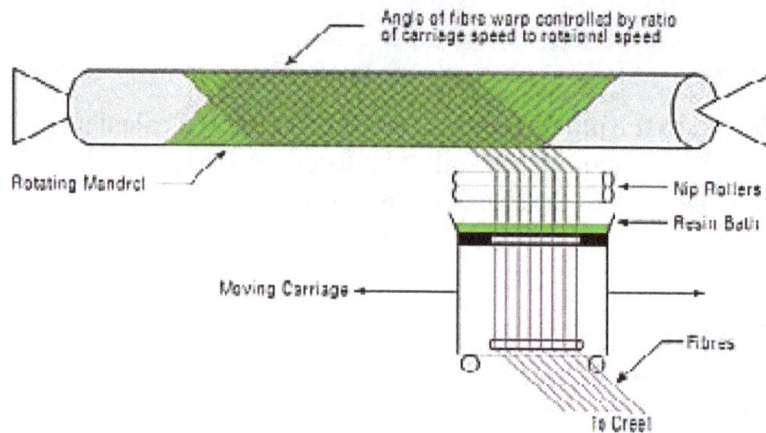

Filament winding

This method can be automated and provides high production rates. Highest-strength products are obtained because of fiber placement control. Control of strength in different direction is also possible.

Matrix Application Methods

The important purpose of the matrix is to bind the filament together and convert into a solid material. The matrix should be free from voids and dirt particles. The matrix application can be divided into seven methods as given below:

Wet Winding

In the wet winding process the matrix in liquid form is placed in a resin bath and the fibers are dipped in that bath and wound. The matrix will be in the liquid form or is brought to liquid form by making a solution. Solids like thermoplastics can be brought to liquid form by melting also.

Dry Winding and Liquid Infiltration

In this process the fibers are initially wound without the matrix and after winding the matrix is allowed to infiltrate into the fibers by pressure injection or vacuum impregnation. The viscosity has to be very low for such impregnations.

Dry Winding and Vapour Infiltration

The fibers are wound first and the matrix is deposited by Chemical Vapour Deposition (CVD) or Physical Vapour Deposition (PVD).

Powder Injection

In this method the matrix material is in the form of powder and is injected into the fibers during winding and then it is converted into solid by sintering or melting.

Prepreg Winding

Prepregs are the fibers, tapes or clothes previously impregnated with the resin. The prepreg will have the fiber to resin ratio correctly maintained. Since the prepregs are in semi solids, winding is more convenient and after winding it is converted into solids by heating or sintering.

Winding of Woven or Comingled Fabrics

In this method the woven fabrics are made by weaving the fiber and the resin which is in the form of fiber. Bundles of reinforcement and matrix filaments are taken as warp and weft fibers and are woven. In the comingled type each bundle of filament will have the matrix and the reinforcement fibers. After winding, the materials are heated to melt the matrix filaments.

Plasma Spraying

The fibers are wound and simultaneously the matrices can be sprayed into it by plasma spraying.

Conversion of Matrix Into Solids

This conversion process can be broadly divided into two processes as below.

Reactive Process

In the reactive process the matrix is formed by a chemical reaction which may be a polymerization process as in Reaction Injection Moulding (RIM) or cross-linking process. In the case of carbon-carbon composites, the infiltrated phenolic is converted into graphite by the polymer pyrolysis.

Reaction Injection Winding

In this process the monomers of the polymeric is infiltrated into the filaments. The reaction continues in-situ converting the monomer into a polymer.

Polymer Cross-Linking

Thermoset resins like epoxies, polyesters, phenolics can be made into liquid linear polymer form. Winding is then done with the resins adding the cross-linking additives. The cross-linked operation continued during winding and after winding the liquid linear polymer is converted into cross-linked solid material.

Non-Reactive Process

Some of these conversions can be described as follows:

Melt Processing

Prepregs are first made by coating the fibers with thermoplastics. The prepregs are then used for winding and after winding the thermoplastics are melted and fused into solids. The prepregs can be made by solution impregnation, melt impregnation, film or powder coating etc.

Sintering

In this process the matrix is in powder form. The fibers are pre-coated with the powder before winding and during winding process the powder is incorporated into fiber by infiltration or powder injection. The powder is subsequently sintered into solid by heating. This method is suitable for materials like polytetra fluoro ethylene (PTFE) which requires very high temperature to melt.

Vapour Deposition

Vapours of metals, ceramics, carbon etc. can be infiltrated into the fiber which is then cooled to form the matrix.

Advantages of Filament Winding

1. Filament winding is semi-automated which can be done more neatly with less workers.

2. Filament winding can give a fiber content as high as 70% by weight in the case of glass fiber.

3. This process is used for making large products like storage tanks up to 15 or 16 m diameter by using special winding machines.

4. It is possible to vary the strength of the wound product in different directions by varying the angle of winding.

Limitations

1. Products with complicated profiles and reverse curvature cannot be wound.

2. The inter-laminar shear strength of the product is low.

3. The ultimate bearing strength is low. They are rigid but less ductile.

4. The laminate quality of the filament wound product is generally lower than that of the product made by autoclave processing.

Materials

Reinforcement Fibers

Glass fiber is the common reinforcement fiber for commercial applications like chemical tanks, petroleum tanks, pipe lines etc. Aramid fibers including Kevlar 49, 29, and 149 are used for making products such as aerospace structures, rocket motor casing etc. Kevlar fibers have poor compressive and shear strength and are not usable for high temperature since they melt at 140°C. Carbon is the next versatile fiber because of their high modulus, strength and temperature resistance. Natural fibers provide good strength for applications like boats and silos by winding process but further research is needed to improve their durability.

Thermoset Resins

Polyester, vinylester, and epoxies are commonly used as thermosets. Polyester including isophthalic and bisphenol resins are used for chemical plants, petroleum tank and pipeline applications. Epoxies, because of their superior shear strength and mechanical and electrical properties used for high performance applications like aerospace and electrical insulation products. Vinylester finds applications in chemical resistant product. Polymide, silicones, phenolics and furan resins finds applications in very special requirements like high temperature resistances.

Thermosets are used in wet winding, prepregs and wet rerolled rovings. Wet rerolled rovings are rovings impregnated with resin and rolled in to spool form and stored under low temperature and then unwound and used for winding.

Thermoplastic Resins

These resins are used to make prepregs by coating the thermoplastics on the fiber by melt dip coating, fiber transfer or by powder coating. The fiber reinforced thermoplastic prepreg tapes are then wound using a tape winding machine. After winding the product can be heated to a level at which the resin melts and fuses into a solid.

Winding Facility

Mandrels

The mandrel constitutes the important part of the winding setup. It is the tool around which the matrix impregnated rovings are wound. The profile of the mandrel gives the profile of the filament wound product. The mandrel must be smooth and easily removable after the product is fully cured.

Mandrels are broadly classified into (i) open ended non-collapsible mandrels and (ii) collapsible mandrels.

Open Ended Non-collapsible Mandrels

They are generally made of steel with smooth surface finishes and an axial taper of 1:200 for easy release of the product from mould. Screw and hydraulic extractors are used for the release of the product.

Collapsible Mandrels

    a.   Segmented metallic collapsible mandrels

The mandrel is made of several segments. The segments are dismantled to release the product.

    b.   Water soluble mandrels

Water soluble mandrels are made by casting over a centered axis and polar fittings, sand and water soluble polyvinyl alcohol.

    c.   Spider plastic mandrels

Plaster of Paris layer is made over removable or collapsible tooling. Plaster can be finished with either duco putty or with release films. After winding central mandrel is removed and the plaster is chopped off.

d.   Inflatable mandrel

Mandrel is made by inflating a bag. They also present the problem of larger transmission.

e.   Low melting alloy mandrels

Low melting alloys and metals like lead can be used for making the mandrel. Later, the mandrel material cannot be recovered.

f.   Non-removable liners

Liners can be made from metals or plastics or FRP for liquid resistant surfaces.

Winding Machines

The winding machine has facilities for wetting the fiber, tensioning the filament, laying the fiber or tapes in the required angle in a uniformly spreaded pattern. Winding machines can be broadly divided into three groups.

a. Helical winding machines

b. Polar winding machines

c. Special purpose and advance winding machines

Helical Winding Machine

The helical winding machine is designed to lay the fiber on a rotating mandrel at winding angles varying from 0°(axial) to 90°(hoop) with axis of rotation. The basic movement of the helical winder is the mandrel rotation and the feed traverse. By varying the speed of two movements, it is possible to vary the winding angle. The feed eye moves to and fro from one end to other end creating an angle ply or netting structure on the mandrel surface.

The fiber is fed to the feed eye through a resin bath in the wet winding process. The resin bath also moves along with the feed eye. In case of prepreg winding the fibers or tapes are fed from a spool or creed stand. The creed stand is stationary or it is fixed with the resin carriage so that the stand also moves along with the feed eye. Figure shows the layout of a typical helical winding machine.

Helical winding machines can be made with constant helix angles in which fibers can be wound only at constant angle with the axis or variable angle machines where the angle of winding can be varied from 0° to 90° with the axis. The variation is achieved by varying the mandrel surface speed and the feed point speed. Accurate speed variations can be possible by using numerically controlled step motors with or without servo hydraulic pulse motors. These machines have the advantage that winding angle can be changed along the length by pre-programmed using punched tapes or by using computers.

Polar Winding Machine

Polar winding is done generally for spherical, ellipsoidal or other closed axis symmetric shells. The two ends of the mandrel is called poles. The winding is done from one pole to the other. Figure shows a typical polar winding machine.

The polar winding machine can be made in two different ways:

1. The feeding eye is rotating while the mandrel is on a fixed axis with only rotating motion. This system needs the resin bath and fiber spool to travel with the feed eye.

2. The feeding eye is fixed while the mandrel has two motions with rotation about its axis and a rotation about one of the mounting supports. The advantage in this is the resin bath need not travel around the mandrel, but the rotation of heavy mandrel has to be done using a cantilever arrangement and the support system must be rigid enough to carry the load without causing any deflections.

Depending upon the way mandrel is supported, the winding machines can be classified into cantilever type or with both ends supported. In the cantilever type, the mandrel is supported at one end and the other end is free which helps to take round the mandrel without any obstruction.

Combined Polar and Helical Winding Machine

Products like pressure vessels, road tankers, petroleum tanks etc. require a cylindrical shell with end domes having spherical, ellipsoidal shapes. One way is to make them separately and joined them together, which gives a weak joint at the junction. The better way is winding the shell and domed ends using combined helical and polar winding machines.

Special Purpose Machines

Several other variations of filament winding machines have been developed for specific end uses and a few machines are described below.

Fixed Mandrel Machines

For winding very large cylindrical tanks, rotating the mandrel for winding becomes very costly. In such cases the mandrel is kept stationary on a vertical axis. The resin bath together with fiber creels move around the mandrel and up and down to create the helical path around the mandrel

Race Track Machines

In these machines the resin bath and fiber creel travels on a race track and the mandrel rotates about its axis. The winding angle is achieved by tilting the mandrel to the required positions.

Continuous Pipe Making Machines

These machines have stationary mandrels. The fiber spools are mounted on a circular ring which rotates around the mandrel. Two such rings rotating in the opposite direction create the helical or angle ply pattern on the mandrel. By controlling the linear and rotary motions of the ring the required winding angle can be obtained.

Braiding Machine

It is similar to the continuous filament winding except that the fibers get knitted during the winding process. Braiding without resin is used for fiber insulation of electrically conductive wires. Braiding with resin is used for making filament wound high pressure hoses.

Resin Curing System

In normal machines, the mandrel with wet wound shell is transferred to an oven where it is cured as per the cure schedule. In order to present the resin dripping, the mandrel is rotated slowly in the chamber until the resin gels. The rotation must be slow to prevent the resin coming to surface due to the centrifugal force. After curing the product is removed and then post cured if necessary.

Another way to cure is to use infrared rays which will be focused on to the wet wound shell. As the mandrel rotates slowly, the radiated heat helps to cure the resin.

For cold curing system, the cure can be achieved by keeping the mandrel rotated slowly until the resin gels. After curing the product will be released and it is post cured if necessary.

For large products, curing the product in oven is expensive. Hence, such products can be post cured using hot air circulation, local heating etc.

Mandrel Extraction Facility

The extraction of the mandrel from the product is a difficult task in many cases particularly when the product is very large in size and also when the resin has a high cure shrinkage. The mandrel extraction can be done by different ways.

1. A hydraulic extractor can be used to pull out the mandrel.

2. The mandrel can be dismantled into pieces and pulled out through the side opening.

3. Dissolving of mandrel.

Spray-up Molding

Spray-up molding is an open mold method that can produce complex parts more economically than hand lay-up. Chopped fiberglass reinforcement and catalyzed resin, and in some cases, filler materials, are deposited on the mold surface from a combination chopper/spray gun.

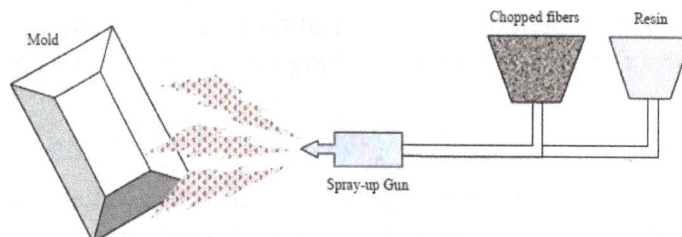

Spray-up molding

Rollers or squeegees are used to manually remove entrapped air and work the resin into the reinforcements. Woven fabric or woven roving is often added in specific areas for greater strength. As in hand lay-up, gel coats are used to produce a high quality colored part surface.

Vacuum Bag Molding

Vacuum impregnation is a process in which the resin to fiber wetting is assisted by a vacuum. The main purpose of vacuum process is it removes the air which is trapped inside the laminate thus reducing the defect and improving the strength of the laminate.

Advantages of Vacuum Impregnation Method

- Improves the strength of the laminates by reducing the defects.
- Low cost when compared with compression mould laminates.
- Density of laminate is considerably lower than that of compression mould laminates.

Materials

Reinforcements

All types of fiber reinforcements can be impregnated with resin using vacuum method.

Effects of vacuum method on reinforcements:

- Good formability.
- High strength.
- Good surface quality
- Wear resistance.
- High complex forms.

Resins

The curing procedure of the resin, initial viscosity, the gelation time, and wettability, are the important properties to be considered for processing. For vacuum impregnation purpose the volatile content should be as low as possible. Both polyester and epoxy resins are used.

Factors to be Considered for Resins in Vacuum Method

- Long pot life
- Less viscosity (11 Pa.s or 100 CP or less)
- Short gel time (less than 1 hour)

Mold Releasing Agents

Releasing agents used include backed on Teflon (PTFE) or PVA coatings on the mould parts. For vacuum bag systems PTFE films which are porous are used along with conventional materials.

Adaptation of Vacuum Impregnation Method

Vacuum impregnation process is used in many related FRP fabrication processes such as:

- Vacuum impregnation
- Vacuum injection moulding
- Vacuum bag moulding

Vacuum Impregnation

Vacuum impregnation is used for the manufacture of products which need precisely controlled mechanical properties, thermal and electrical stability and good dimensional control.

Mould surface is treated with releasing agent. Reinforcements are then placed inside the mould. While closing the mould, care has to be taken to see whether it is completely sealed. Otherwise, when vacuum is applied to the mould, full vacuum may not be generated in the mould cavities.

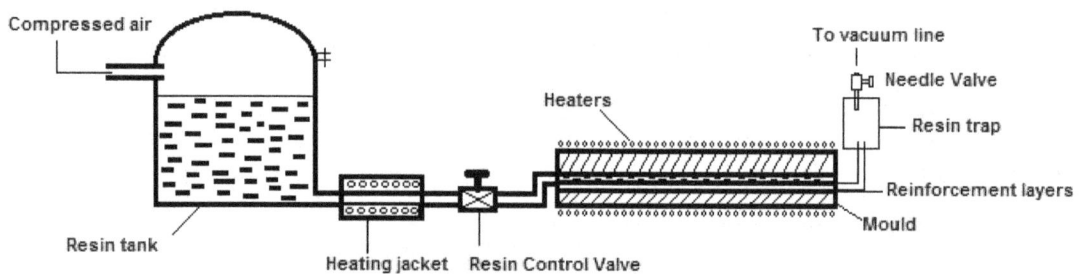

Vacuum impregnation

Vacuum is applied so that the resin gradually fills up the mould cavity and wets the reinforcements. Once the resin is completely filled in, the heating can be applied to accelerate the gelation and curing of the resin. After cure, the mould is declamped and the product is taken out.

Vacuum Injection Moulding

It is a process where combination of vacuum impregnation and resin injection system are adapted. It is also known as Hoechst process. Moulds for this process can be made of GRP, lower half is of a rigid construction while the upper half is more flexible. A vacuum channel is built into the mould around the periphery for mould closure.

As usual the mould surfaces are waxed, polished and coated with PVA release agent. The reinforcements are cut to shape and fitted in the lower half of the mould. Once the upper half in position, vacuum is applied on the gasket channel sealing the mould and renders it air tight.

Catalyzed resin is injected under pressure. The air remaining in the mould is sucked out, while the flexible top half forces the resin to flow through the reinforcements until the mat is thoroughly impregnated and compacted.

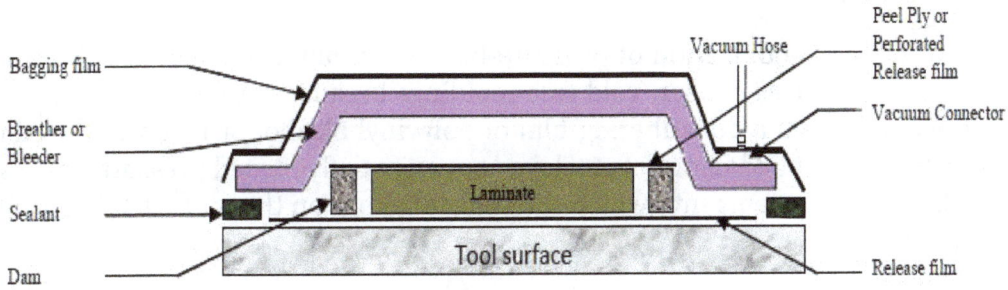

Vacuum Bag method

The lay-up is placed in the mould with separator on top of it usually made of Teflon coated glass. The bleeder is placed to absorb the excess resin from the lay-up, thus controlling the amount content during the curing process. Pressure plates are introduced to supply additional compression to the lay-up. Barrier film is used to control the resin flow in the bleeder.

Breather film is given beneath the vacuum bag to allow the uniform application of vacuum all over the area of the laminate and removal of excess air or volatiles developed during the cure.

Vacuum bag is used to contain the vacuum generated by the pump and applied to the lay-up. The application of vacuum bag is very critical. Bag porosity or punctures can result in a porous product. Complex tools may require the bag to be folded in places and thus require excess bag material. If the folds are not properly made or placed, wrinkles may be developed in the parts. The vacuum may be maintained till the resin gels.

Pressure Bag Molding

Pressure bag molding is similar to the vacuum bag molding method except that air pressure, usually 200 to 350 kPa, is applied to a rubber bag, or sheet that covers the laid up composite to force out entrapped air and excess resin. Pressurized steam may be used instead, to accelerate the cure. Cores and inserts can be used with the process, and undercuts are practical, but only female and split molds can be used to make items such as tanks, containers, and wind turbine blades.

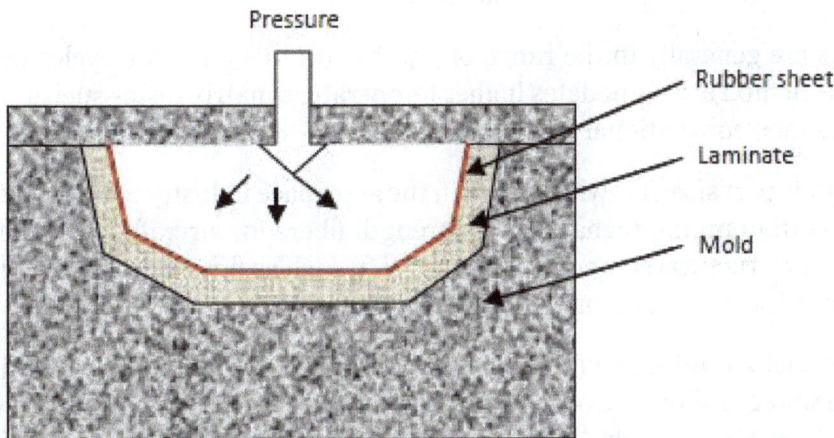

Pressure bag molding

Autoclave Molding

Autoclave molding is a modification of pressure-bag and vacuum-bag molding. This advanced composite process produces denser, void free moldings because higher heat and pressure are used for curing. and resin, a nonadhering film of polyvinyl alcohol or nylon is placed over the lay-up and sealed at the mold flange. Autoclaves are essentially heated pressure vessels usually equipped with vacuum systems into which the bagged lay-up on the mold is taken for the cure cycle.

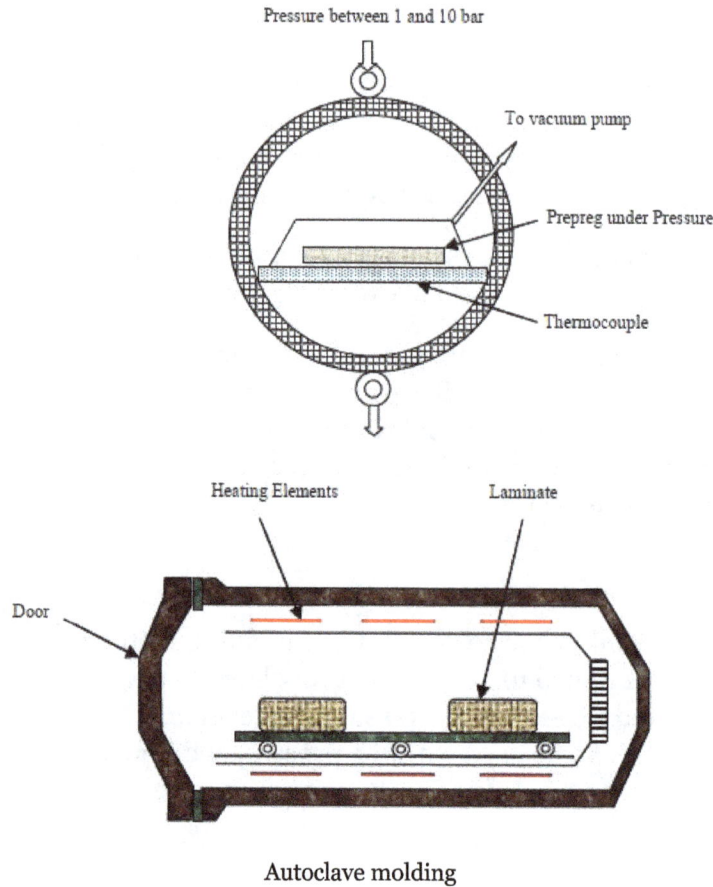

Autoclave molding

Curing pressures are generally in the range of 350 to 700 kPa and cure cycles normally involve many hours. The method accommodates higher temperature matrix resins such as epoxies, having higher properties than conventional resins.

Autoclave size limits part size. It is widely used in the aerospace industry to fabricate high strength/weight ratio parts from preimpregnated high strength fibers for aircraft, spacecraft and missiles. Many large primary structural components for aircraft, such as fins, wing spars and skins, fuselages and flying control surfaces, are manufactured by this method.

The starting material for autoclave moulding process is prepreg. A prepreg contains 42% weight of resin. If this prepreg is allowed to cure without any resin loss the cures laminate would contains 50% volume of fibers. Since, nearly 10% weight of resin flows out during the moulding process, the actual volume of fiber in the cured laminate is 60%.

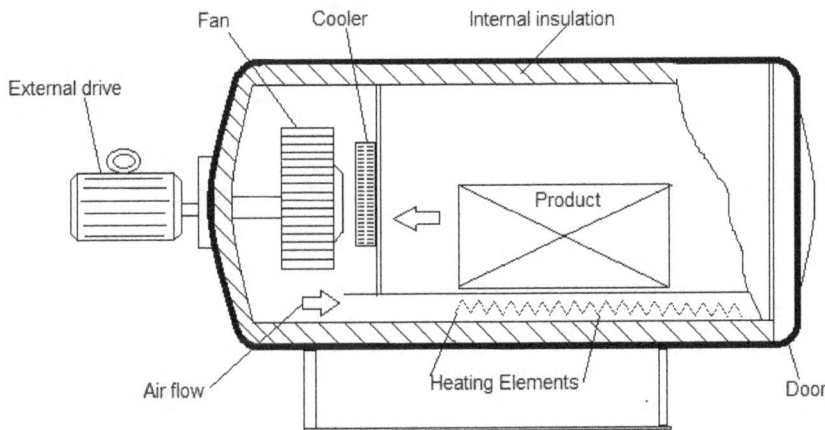

(a) Autoclave Setup

After layup, a porous release cloth and a few layers of bleeder papers are placed on top of the prepreg stack. The bleeder paper is used to absorb the excess resin in the moulding process. The complete layup is covered with another Teflon sheet and then a thin heat resistant vacuum bag. The entire assembly is kept inside a preheated autoclave where a combination of pressure and temperature is applied and the plies are converted into a solid laminate.

As the prepreg is heated in the autoclave, the resin viscosity in the B-stage prepreg decreases up to its minimum and then increases rapidly as the curing reaction begins.

Cure cycle of an epoxy prepreg consists of two stages:

The first stage consists of increasing the temperature up to 130°C and dwelling at this temperature for 60 min. When the minimum viscosity reaches external pressure is applied to flow out the excess resin into the bleeder papers. This will remove the air entrapment and volatile from the prepreg.

At the end of temperature dwell, the autoclave temperature resets to the actual curing temperature of the resin. The cure temperature and the pressure is maintained for 2 hours or complete cure takes place. At the end of the cure cycle, the temperature is slowly reduced while the laminate is still under pressure. Finally, the laminate is removed from the bag and post cured if needed.

Equipment

The following data have to be specified for autoclave:

1) Maximum operating temperature

2) Rate of temperature rise

3) Rate of temperature decrease

4) Temperature control stability

5) Stabilized temperature uniformity

6) Maximum pressure

7) Pressurizing medium

8) Pressurization

9) Depressurization

10) Number of vacuum stations

11) Maximum exterior surface temperature

12) Workspace size

13) Heating

14) Cooling

Bagging Materials, Release Sheets, Peel Plies and Breather Cloths

Bagging: Applying an impermeable layer of thin film over an uncured part and sealing edges so that a vacuum can be drawn.

Bagging film sealant tape: This is a soft mastic type of tape which is slightly tacky and is used to seal bagging film.

Breather cloth: A loosely woven material such as a glass fabric that will serve as a continuous vacuum path but not direct contact with the part. Its purpose is to allow removal of air, thereby applying atmospheric pressure to the part.

Bleeder cloth: A non-structural layer of material, used to allow the escape of gas and excess resin during the cure. The bleeder cloth is removed after the curing.

Peel ply: A layer of open-weave material, applied directly to the surface of a prepreg layup. Peel ply is removed from the cured laminate.

Release film: A material of thin film, used to keep the resin from bonding to the mould. Release films are made from non stick materials such as polyvinyl fluoride (PVF), fluorinated ethylene propylene (FEP), polyester and nylon.

Other Techniques

There are few techniques which are recently developed for making composite products, This includes :

➢ Tube rolling

➢ Elastic reservoir molding

➢ Resin film infusion

➢ Reaction Injection Molding (RIM)

➢ Structural reaction injection molding

References

- Shaffer, G.D. "An Archaeomagnetic Study of a Wattle and Daub Building Collapse." Journal of Field Archaeology, 20, No. 1. Spring, 1993. 59-75. JSTOR. Accessed 28 January 2007

- Khurram, Shehzad; Xu, Yang; Chao, Gao; Xianfeng, Duan (2016). "Three-dimensional macro-structures of two-dimensional nanomaterials". Chemical Society Reviews. doi:10.1039/C6CS00218H

- "Minerals commodity summary – cement – 2007". US United States Geological Survey. 1 June 2007. Retrieved 16 January 2008

- Kim, Hyoung Seop (2000-09-30). "On the rule of mixtures for the hardness of particle reinforced composites". Materials Science and Engineering: A. 289 (1): 30–33. doi:10.1016/S0921-5093(00)00909-6

- "The Composites Design and Manufacturing HUB". Composites Manufacturing. June 11, 2014. Retrieved September 2, 2016

- Aghdam, M. M.; Morsali, S. R. (2013-11-01). "Damage initiation and collapse behavior of unidirectional metal matrix composites at elevated temperatures". Computational Materials Science. 79: 402–407. doi:10.1016/j.commatsci.2013.06.024

An Integrated Study of Lamination

Lamination is the method of manufacturing materials in layers. This process helps in improving the strength and stability of the material. Lamination is one of the methods that reduce stress and strain to a material, and this aspect is studied in the concept of strength of materials. The aspects elucidated in this chapter are of vital importance, and provide a better understanding of lamination.

Lamination

Lamination is the technique of manufacturing a material in multiple layers, so that the composite material achieves improved strength, stability, sound insulation, appearance or other properties from the use of differing materials. A laminate is a permanently assembled object by heat, pressure, welding, or adhesives.

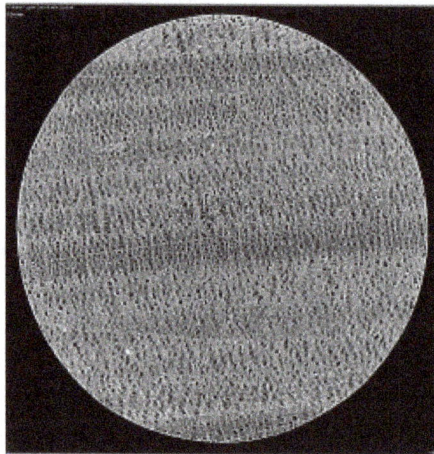

Flight through a µCT image stack of a knitting needle that consists of laminated wooden layers. The layers can be differentiated by the change of direction of the wood´s vessels.

Materials

There are different lamination processes, depending on the type of materials to be laminated. The materials used in laminates can be the same or different, depending on the processes and the object to be laminated. An example of the type of laminate using different materials would be the application of a layer of plastic film—the "laminate"—on either side of a sheet of glass—the *laminated* subject.

Vehicle windshields are commonly made by laminating a tough plastic film between two layers of glass. This is to prevent shards of glass detaching from the windshield in case it breaks. Plywood is a common example of a laminate using the same material in each layer. Glued and laminated dimensioned timber is used in the construction industry to make wooden beams, Glulam, with sizes larger and stronger than can be obtained from single pieces of wood. Another reason to laminate wooden strips into beams is quality control, as with this method each and every strip can be inspected before it becomes part of a highly stressed component.

Building Materials

Examples of laminate materials include melamine adhesive countertop surfacing and plywood. Decorative laminates are produced with decorative papers with a layer of overlay on top of the decorative paper, set before pressing them with thermoprocessing into high-pressure decorative laminates. A new type of HPDL is produced using real wood veneer or multilaminar veneer as top surface. High-pressure laminates consists of laminates "molded and cured at pressures not lower than 1,000 lb per sq in.(70 kg per sq cm) and more commonly in the range of 1,200 to 2,000 lb per sq in. (84 to 140 kg per sq cm). Meanwhile, low pressure laminate is defined as "a plastic laminate molded and cured at pressures in general of 400 pounds per square inch (approximately 27 atmospheres or 2.8×106 pascals).

Paper

Corrugated fiberboard boxes are examples of laminated structures, where an inner core provides rigidity and strength, and the outer layers provide a smooth surface.

Laminating paper products, such as photographs, can prevent them from becoming creased, faded, water damaged, wrinkled, stained, smudged, abraded, or marked by grease or fingerprints. Photo identification cards and credit cards are almost always laminated with plastic film. Boxes and other containers are also laminated using a UV coating. Lamination is also used in sculpture using wood or resin. An example of an artist who used lamination in his work is the American Floyd Shaman.

Further, laminates can be used to add properties to a surface, usually printed paper, that would not have them otherwise. Sheets of vinyl impregnated with ferro-magnetic material can allow portable printed images to bond to magnets, such as for a custom bulletin board or a visual presentation. Specially surfaced plastic sheets can be laminated over a printed image to allow them to be safely written upon, such as with dry erase markers or chalk. Multiple translucent printed images may be laminated in layers to achieve certain visual effects or to hold holographic images. Many printing businesses that do commercial lamination keep a variety of laminates on hand, as the process for bonding many types is generally similar when working with arbitrarily thin material.

Photo Laminators

Three types of laminators are used most often in digital imaging:

- Pouch laminators

- Heated roll laminators

- Cold roll laminators

Film Types

Laminate film is generally categorized into these five categories:

- Standard thermal laminating films
- Low-temperature thermal laminating films
- Heat set (or heat-assisted) laminating films
- Pressure-sensitive films
- Liquid laminate

Micro-mechanics of Lamina

Micromechanics deals with the study of composite material behaviour in terms of the interaction of its constituents. From the procedures of micromechanics lamina properties can be predicted. There are two basic approaches of the micromechanics of composite materials, namely

(i) Mechanics of materials and (ii) Elasticity

Here, the mechanics of materials approach will be followed.

Volume Fractions

Consider a composite material that consists of fibers and matrix material. The volume of the composite material is equal to the sum of the volume of the fibers and the volume of the matrix. Therefore,

$$v_c = v_f + v_m$$

where, v_c - volume of composite material

v_f - volume of fiber

v_m - volume of matrix

Let, the fiber volume fraction V_f and the matrix volume fraction V_m be defined as

$$V_f = \frac{v_f}{v_c} \text{ and}$$

$$V_m = \frac{v_m}{v_c}$$

such that the sum of volume fractions is

$$V_f + V_m = 1$$

Weight Fractions

Assuming that the composite material consists of fibers and matrix material, the weight of the composite material is equal to the sum of the weight of the fibers and the weight of the matrix. Therefore,

$$w_c = w_f + w_m$$

where, w_c - weight of composite material

w_f - weight of fiber

w_m - weight of matrix

The weight fractions (mass fractions) of the fiber and the matrix are defined as

$$W_f = \frac{W_f}{W_c}$$ and

$$W_m = \frac{W_m}{W_c}$$

such that the sum of weight fractions is

$$W_f + W_m = 1$$

Density

The density of composite material can be defined as the ratio of weight of the composite material to the volume of the composite material and is expressed as

$$\rho_c = \frac{w_c}{v_c}$$

But, $v_c = v_f + v_m$, and $v = \frac{w}{\rho}$, therefore the above equation can be rewritten as

$$\frac{w_c}{\rho_c} = \frac{w_f}{\rho_f} + \frac{w_m}{\rho_m}$$

$$\frac{w_c}{\rho_c} = \frac{w_f}{\rho_f} + \frac{w_m}{\rho_m}$$

$$\frac{1}{\rho_c} = \frac{1}{\rho_f}\left(\frac{w_f}{w_c}\right) + \frac{1}{\rho_m}\left(\frac{w_m}{w_c}\right)$$

By writing in terms of weight fractions,

$$\frac{1}{\rho_c} = \frac{W_f}{\rho_f} + \frac{W_m}{\rho_m}$$

$$\frac{1}{\rho_c} = \frac{W_f}{\rho_f} + \frac{W_m}{\rho_m}$$

The density of the composite material in terms of weight fractions can be written as

$$\rho_c = \frac{1}{\left(\dfrac{W_f}{\rho_f} + \dfrac{W_m}{\rho_m}\right)}$$

$$\rho_c = \frac{1}{\sum_{i=1}^{n}\left(\dfrac{W_i}{\rho_i}\right)}$$

Moreover, the equation $w_c = w_f + w_m$, in general, can be rewritten as

$$\rho_c v_c = \rho_f v_f + \rho_m v_m$$

$$\rho_c = \rho_f\left(\frac{v_f}{v_c}\right) + \rho_m\left(\frac{v_f}{v_c}\right)$$

writing in terms of volume fractions, the density of the composite material is written as

$$\rho_c = \rho_f V_f + \rho_m V_m$$

In general,

$$\rho_c = \sum_{i=1}^{n} \rho_i V_i$$

Void Content

During the incorporation of fibers into the matrix or during the manufacturing of laminates, air or other volatiles may be trapped in the material. The trapped air or volatiles exist in the laminate as micro voids, which may significantly affect some of its mechanical properties. A high void content (over 5% by volume) usually leads to lower fatigue resistance, greater susceptibility to water diffusion, and increased variation (scatter) in mechanical properties. The void content in a composite laminate can be estimated by comparing the theoretical density with its actual density.

$$v_{void} = \left(\frac{\rho_{ct} - \rho_{ce}}{\rho_{ct}}\right) * 100$$

where, ρ_{ct} - theoretical density of the composite material

ρ_{ce} - experimental density of the composite material

Determination of Longitudinal Modulus

Consider a unidirectional composite specimen as shown in Figure.

The following assumptions are made to get the basic properties:

    (i) Fibers are uniform in properties and diameter.

    (ii) Fibers are continuous and parallel throughout the composites.

    (iii) There is a perfect bonding between the fibers and the matrix.

    (iv) Strains experienced by the fiber, matrix and composites are equal. i.e.

$$\varepsilon_c = \varepsilon_f = \varepsilon_m$$

where, $\varepsilon_c = \varepsilon_f$ and ε_m are the longitudinal strains in fibers, matrix, and composite respectively. This condition is called iso-strain condition.

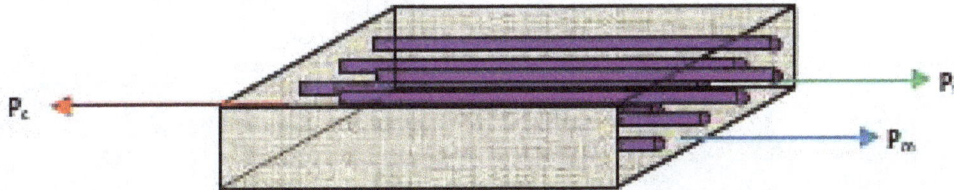

Model for longitudinal behaviour of composite material

Let, the composite be applied by a load Pc which is shared between the fibers and the matrix so that

$$P_c = P_f + P_m$$

The corresponding stress relation is

$$\sigma_c A_c = \sigma_f A_f + \sigma_m A_m$$

Further its behaviour is assumed to be linearly elastic, and hence the modulus and stress are related. Thus,

$$E_c \varepsilon_c A_c = E_f \varepsilon_f A_f + E_m \varepsilon_m A_m$$

$$E_c = E_f \left(\frac{\varepsilon_f A_f}{\varepsilon_c A_c} \right) + E_m \left(\frac{\varepsilon_m A_m}{\varepsilon_c A_c} \right)$$

But, for parallel fibers the area fraction is same as the volume fraction. Thus,

$$E_c = E_f V_f + E_m V_m$$

The relationship of this form is known as *Rule or Law of Mixtures*

$$E_c = E_f V_f + E_m \left(1 - V_f\right)$$

In general

$$E_c = \sum_{i=1}^{n} E_i V_i$$

Longitudinal Strength

The load is shared by the fibers and the matrix.

$$P_c = P_f + P_m$$

Thus,

$$\sigma_c A_c = \sigma_f A_f + \sigma_m A_m$$

$$\sigma_c = \sigma_f \left(\frac{A_f}{A_c}\right) + \sigma_m \left(\frac{A_m}{A_c}\right)$$

$$\sigma_c = \sigma_f V_f + \sigma_m V_m$$

$$\sigma_c = \sigma_f V_f + \sigma_m \left(1 - V_f\right)$$

Load Carrying Capacity of Fibers

The strains experienced by the composites, fibers and the matrix are equal. Thus,

$$\varepsilon_c = \varepsilon_f = \varepsilon_m$$

$$\frac{\sigma_c}{E_c} = \frac{\sigma_f}{E_f} = \frac{\sigma_m}{E_m}$$

$$\frac{\sigma_f}{\sigma_m} = \frac{E_f}{E_m} \ and \ \frac{\sigma_f}{\sigma_c} = \frac{E_f}{E_c}$$

$$\frac{P_f}{P_m} = \frac{\sigma_f A_f}{\sigma_m A_m} = \frac{\dfrac{\left(\llbracket E_f \varepsilon_f \right) A \rrbracket_f}{Ac}}{\dfrac{\left(E_m \varepsilon_m\right) A_m}{Ac}} = \frac{E_f V_f}{E_m V_m}$$

$$\frac{P_f}{P_c} = \frac{\sigma_f A_f}{\sigma_f A_f + \sigma_m A_m}$$

$$\frac{P_f}{P_c} = \frac{\sigma_f A_f \frac{A_m}{Ac}}{\frac{[(\sigma)]_f A_f + \sigma_m A_m}{Ac}} = \frac{\sigma_f V_f}{\sigma_f V_f + \sigma_m V_m}$$

$$\frac{P_f}{P_c} = \frac{E_f \varepsilon_f V_f}{E_f \varepsilon_f V_f + E_m \varepsilon_m V_m}$$

$$\frac{P_f}{P_c} = \frac{E_f V_f}{E_f V_f + E_m V_m}$$

$$\text{or,} \quad \frac{P_f}{P_c} = \frac{\dfrac{E_f}{E_m}}{\dfrac{E_f}{E_m} + \dfrac{V_m}{V_f}}$$

Thus, the load sharing of the fibers depend on the modulus values and the volume fractions of fiber and matrix.

Determination of Transverse Modulus

Consider unidirectional composites. The following assumptions are made

(i)Fibers are uniform in properties and diameter.

(ii)Fibers are continuous and parallel throughout the composite.

(iii)There is a perfect bonding between the fibers and the matrix.

(iv)The fibers and the matrix are made up of layers and each layer will carry the same load. Therefore the fiber and matrix layers will experience equal stress. i.e.

$$\sigma_c = \sigma_f = \sigma_m$$

where, σ_c, σ_f and σ_m are the stresses in the composites, fibers and matrix respectively, in the loading direction (transverse direction).

(v) The thickness of the composite material is equal to the sum of the thickness

of fibers and matrix. i.e.

$$t_c = t_f + t_m$$

where, t_c, t_f and t_m are the thicknesses of composites, fibers, and matrix, and respectively

Let, the load be applied in the transverse direction, i.e. the direction perpendicular to the parallel fibers.

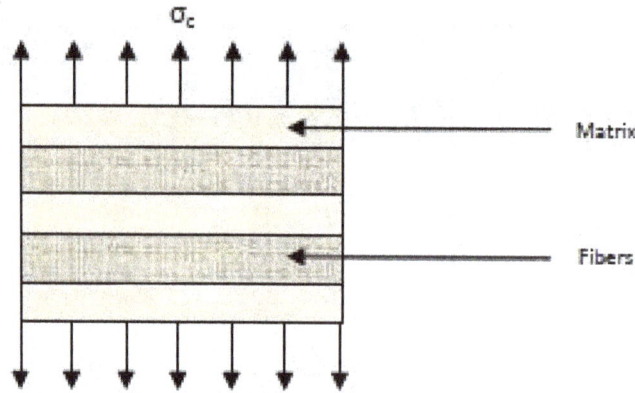

Model for transverse behaviour of composite material

Under the applied transverse load, the elongation of composite material δc in the direction of the load is the sum of the fiber elongation and the matrix elongation. i.e.

$$\delta_c = \delta_f + \delta_m$$

where, δ_c, δ_f and δ_m are the elongations of composite, fibers, and

matrix respectively

but,   strain, $\varepsilon = \delta / t$

and it gives $\delta = \varepsilon * t$

therefore, the above equation can be rewritten as

$$\varepsilon_c t_c = \varepsilon_f t_f + \varepsilon_m t_m$$

$$\varepsilon_c = \varepsilon_f \left(\frac{t_f}{t_c} \right) + \varepsilon_m \left(\frac{t_m}{c} \right)$$

$$\varepsilon_c = \varepsilon_f V_f + \varepsilon_m V_m$$

$$\frac{\sigma_c}{E_c} = \frac{\sigma_f}{E_f} V_f + \frac{\sigma_m}{E_m} V_m$$

But, $\sigma_c = \sigma_f = \sigma_m$

Therefore,

$$\frac{1}{E_{cT}} = \frac{V_f}{E_f} + \frac{V_m}{E_m}$$

$$E_{cT} = \frac{E_f * E_m}{E_f V_m + E_m V_f}$$

or $\quad E_{cT} = \dfrac{1}{\sum_{i=1}^{n} \dfrac{V_i}{E_i}}$

Theories of Stress Transfer in Short Fibers

Tensile load applied to a discontinuous fiber lamina is transferred to the fibers by a shearing mechanism between fibers and matrix. Since, the matrix has low modulus, the longitudinal strain in the matrix is higher than that in the adjacent fibers. If a perfect bond is assumed between the two constituents, the difference in longitudinal strains creates a shear stress distribution across the fiber–matrix interface. Ignoring the stress transfer at the fiber end cross sections and the interaction between the neighboring fibers, we can calculate the normal stress distribution in a discontinuous fiber by a simple force equilibrium analysis.

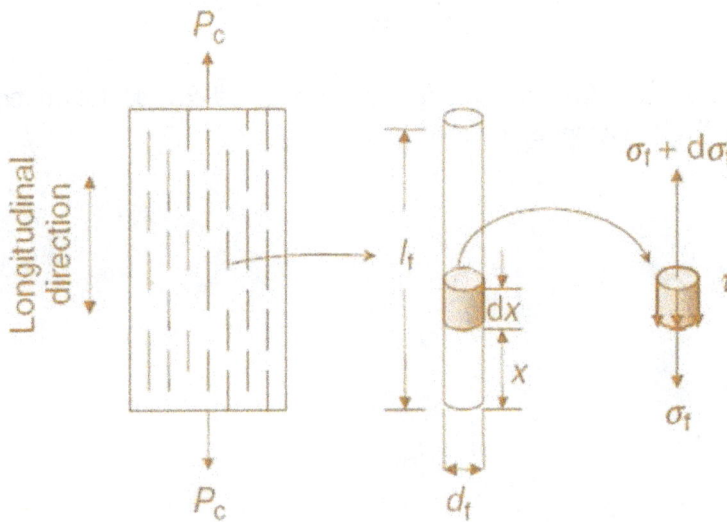

Stress distribution on short fiber

Consider an infinitesimal length dx at a distance x from one of the fiber ends. The force equilibrium equation for this length is

$$\left(\pi r^2\right)\sigma_f + \left(2\pi r\, dz\right)\tau = \left(\pi r^2\right)\left(\sigma_f + d\sigma_f\right)$$

which on simplification gives

$$\frac{d\sigma_f}{dz} = \frac{2\tau}{r}$$

Where,

σ_f is the fiber stress in the axial direction

τ is the shear stress on the cylindrical fiber-matrix interface

r is the fiber radius

$$\sigma_f = \sigma_{fo} + \frac{2}{r}\int_0^z \tau\, dz$$

where, σ_{fo} is the stress on the fiber end. In many analyses σ_{fo} is neglected because of yielding of the matrix adjacent to the fiber end or separation of the fiber end from the matrix as a result of large stress concentrations. Therefore,

$$\sigma_f = \frac{2}{r}\int_0^z \tau\, dz = \frac{2\tau_z}{r}$$

The maximum fiber stress occurs at the midfiber length $i.e.$, at $z = \frac{1}{2}$

$$\left(\sigma_f\right)_{max} = \frac{\tau l}{r}$$

Based on the assumption that the strains in fibers, matrix and composite are equal, the maximum fiber stress $\left(\sigma_f\right)_{max}$ is limited as given below.

$$\varepsilon_c = \varepsilon_f$$

$$\frac{\left(\sigma_f\right)_{max}}{E_f} = \frac{\sigma_c}{E_c}$$

$$\left(\sigma_f\right)_{max} = \frac{E_f}{E_c}\sigma_c$$

The minimum fiber length may be defined as a load-transfer length in which the maximum fiber stress, $\sigma_{f,max}$ can be achieved on the application of the external load, σ_c.

The minimum fiber length, $l_t = \dfrac{\sigma_{f,max}}{\tau}\dfrac{d}{2} = \dfrac{\left(E_f/(E_c)\sigma c^d\right)}{2\tau}$

where, 'd' is the diameter of the fiber

The critical fiber length may be defined as the minimum fiber length in which the maximum allowable fiber stress or the fiber ultimate strength, σ_u can be achieved.

The critical fiber length, $l_c = \dfrac{\sigma_f u}{\tau}\dfrac{d}{2}$

(i) Thus the minimum fiber length, l_t, is based on the applied stress, σ_c, whereas the critical fiber length, l_c, is independent of applied stress, σ_c.

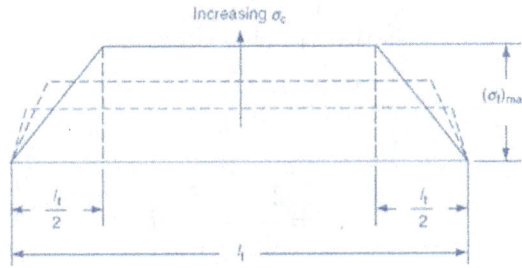

(a) stress on short fiber

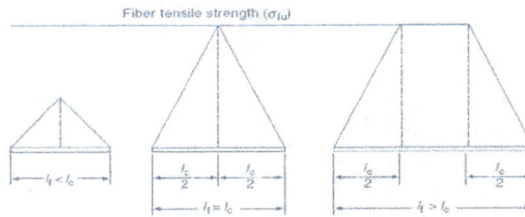

(b) Variations of fiber stress for different fiber lengths

From the figure, the following points are deduced.

(i)   For $l_f < l_c$, the maximum fiber stress may never reach the ultimate fiber strength. In this case, either the fiber–matrix interfacial bond or the matrix may fail before fibers achieve their ultimate strength.

(ii) For $l_f > l_c$, the maximum fiber stress may reach the ultimate fiber strength over much of its length. However, over a distance equal to $l_c/2$ from each end, the fiber remains less effective.

(iii) For effective fiber reinforcement, that is, for using the fiber to its ultimate strength, one must select lf >>lc.

(iv) For a given fiber diameter and strength, l_c can be controlled by increasing or decreasing τ. For example, a matrix-compatible coupling agent may increase τ, which in turn decreases l_c. If l_c can be reduced relative to l_f through proper fiber surface treatments, effective reinforcement can be achieved without changing the fiber length.

Prediction of Modulus of Short Fibers

The Halpin-Tsai equations provide the way to calculate various moduli of aligned short-fiber composites.

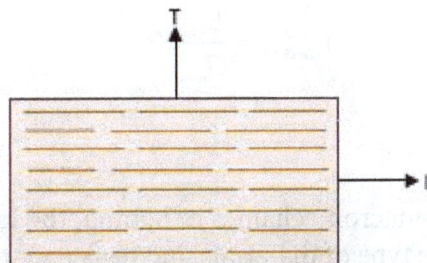

Model of an aligned short-fiber compsite

$$E_L = \frac{1 + 2\left(\dfrac{l_f}{d_f}\right)\eta_L V_f}{1 - \eta_L V_f} E_m$$

$$E_T = \frac{1 + 2\eta_T V_f}{1 - \eta_T V_f} E_m$$

$$G_{LT} = \frac{1 + \eta_G V_f}{1 - \eta_G V_f} G_m$$

$$v_{LT} = V_f v_f + V_m v_m$$

Where,

$$\eta_L = \frac{E_f / E_m - 1}{\dfrac{E_f}{E_m} + 2\left(\dfrac{l_f}{d_f}\right)}$$

$$\eta_T = \frac{E_f / E_m - 1}{E_f / E_m + 2}$$

$$\eta_G = \frac{G_f / G_m - 1}{G_f / G_m + 1}$$

The short fiber composite with random orientation produces the composite with isotropic behaviour in a plane. To predict the elastic moduli of such randomly oriented composites, the empirical formulae given below are used.

$$E_{random} = \frac{3}{8} E_L + \frac{5}{8} E_T$$

$$G_{random} = \frac{1}{8} E_L + \frac{1}{4} E_T$$

$$v_{random} = \frac{E_{random}}{G_{random}} - 1$$

Macro-mechanics of Lamina

For better understanding of the macromechanics of lamina, the knowledge of the material properties in essential. Therefore, the type of materials and tests conducted to find material constants are discussed first.

Types of Materials

The materials are classified based on the behaviour for a particular loading condition. These include,

(i) Anisotropic materials

(ii) Monoclinic materials

(ii) Orthotropic materials

(iii) Transversely isotropic materials

(iv) Isotropic materials

Anisotropic Materials

In an anisotropic material, there are no planes of material property symmetry. So, it has different physical properties in different directions relative to the crystal orientation of the materials i.e., material properties are directionally dependent.

$$\begin{Bmatrix} \sigma_{11} \\ \sigma_{22} \\ \sigma_{33} \\ \tau_{23} \\ \tau_{31} \\ \tau_{12} \end{Bmatrix} = \begin{pmatrix} C_{11} & C_{12} & C_{13} & C_{14} & C_{15} & C_{16} \\ C_{12} & C_{22} & C_{23} & C_{24} & C_{25} & C_{26} \\ C_{13} & C_{23} & C_{33} & C_{34} & C_{35} & C_{36} \\ C_{14} & C_{24} & C_{34} & C_{44} & C_{45} & C_{46} \\ C_{15} & C_{25} & C_{35} & C_{45} & C_{55} & C_{56} \\ C_{16} & C_{26} & C_{36} & C_{46} & C_{56} & C_{66} \end{pmatrix} \begin{Bmatrix} \varepsilon_{11} \\ \varepsilon_{22} \\ \varepsilon_{33} \\ \gamma_{23} \\ \gamma_{31} \\ \gamma_{12} \end{Bmatrix}$$

There are 21 independent elastic constants in the stress-strain relationship as given above.

As shown in the above relations, there are couplings between the stresses and strains. Normal stresses produce not only normal strains in other directions due to Poisson effect but, also shear strains due to the effect of mutual influence. Similarly, shear stresses produce not only shear strains but also normal strains.

In an anisotropic material, a combination of extensional and shear deformation is produced by a normal stress acting in any direction. This phenomenon of creating both extensional and shear deformations by the application of either normal or shear stresses is termed as extension-shear coupling and is not observed in isotropic materials.

Monoclinic Materials

It has a single plane of material property symmetry. If xy plane (i.e.; 1-2 plane) is considered as the plane of material symmetry then, there are 13 independent elastic constants in the stiffness matrix as given below.

$$\begin{Bmatrix} \sigma_{11} \\ \sigma_{22} \\ \sigma_{33} \\ \tau_{23} \\ \tau_{31} \\ \tau_{12} \end{Bmatrix} = \begin{pmatrix} C_{11} & C_{12} & C_{13} & 0 & 0 & C_{16} \\ C_{12} & C_{22} & C_{23} & 0 & 0 & C_{26} \\ C_{13} & C_{23} & C_{33} & 0 & 0 & C_{36} \\ 0 & 0 & 0 & C_{44} & C_{45} & 0 \\ 0 & 0 & 0 & C_{45} & C_{55} & 0 \\ C_{16} & C_{26} & C_{36} & 0 & 0 & C_{66} \end{pmatrix} \begin{Bmatrix} \varepsilon_{11} \\ \varepsilon_{22} \\ \varepsilon_{33} \\ \gamma_{23} \\ \gamma_{31} \\ \gamma_{12} \end{Bmatrix}$$

As there is a single plane of material property symmetry, shear stresses from the planes in which one of the axis is the perpendicular axis of the plane of material symmetry (i.e.; 2-3 and 3-1 planes) will contribute only to the shear strains in those planes. And normal stresses will not contribute any shear strains in these planes.

Orthotropic Materials

There are three mutually orthogonal planes of material property symmetry in an orthotropic material. Fiber-reinforced composites, in general, contain the three orthogonal planes of material property symmetry and are classified as orthotropic materials. The intersections of these three planes of symmetry are called the principal material directions.

The material behaviour is called as specially orthotropic, when the normal stresses are applied in the principal material directions. Otherwise, it is called as general orthotropic which behaves almost equivalent to anisotropic material.

There are nine independent elastic constants in the stiffness matrix as given below for a specially orthotropic material.

$$\begin{Bmatrix} \sigma_{11} \\ \sigma_{22} \\ \sigma_{33} \\ \tau_{23} \\ \tau_{31} \\ \tau_{12} \end{Bmatrix} = \begin{pmatrix} C_{11} & C_{12} & C_{13} & 0 & 0 & 0 \\ C_{12} & C_{22} & C_{23} & 0 & 0 & 0 \\ C_{13} & C_{23} & C_{33} & 0 & 0 & 0 \\ 0 & 0 & 0 & C_{44} & 0 & 0 \\ 0 & 0 & 0 & 0 & C_{55} & 0 \\ 0 & 0 & 0 & 0 & 0 & C_{66} \end{pmatrix} \begin{Bmatrix} \varepsilon_{11} \\ \varepsilon_{22} \\ \varepsilon_{33} \\ \gamma_{23} \\ \gamma_{31} \\ \gamma_{12} \end{Bmatrix}$$

From the stress-strain relationship it is clear that normal stresses applied in one of the principal material directions on an orthotropic material cause elongation in the direction of the applied stresses and contractions in the other two transverse directions. However, normal stresses applied in any directions other than the principal material directions create both extensional and shear deformations.

Transversely Isotropic Materials

If a material has axes of symmetry in its longitudinal axis and all directions perpendicular to its longitudinal axis (i.e., more than three mutually perpendicular axes of symmetry) then such material is transversely isotropic. (e.g., unidirectional composites). There are five independent elastic constants for these materials.

$$
\begin{Bmatrix} \sigma_{11} \\ \sigma_{22} \\ \sigma_{33} \\ \tau_{23} \\ \tau_{31} \\ \tau_{12} \end{Bmatrix} =
\begin{pmatrix}
C_{11} & C_{12} & C_{13} & 0 & 0 & 0 \\
C_{12} & C_{11} & C_{13} & 0 & 0 & 0 \\
C_{13} & C_{13} & C_{33} & 0 & 0 & 0 \\
0 & 0 & 0 & C_{44} & 0 & 0 \\
0 & 0 & 0 & 0 & C_{44} & 0 \\
0 & 0 & 0 & 0 & 0 & \dfrac{(C_{11}-C_{12})}{2}
\end{pmatrix}
\begin{Bmatrix} \varepsilon_{11} \\ \varepsilon_{22} \\ \varepsilon_{33} \\ \gamma_{23} \\ \gamma_{31} \\ \gamma_{12} \end{Bmatrix}
$$

Isotropic Material

In an isotropic material, properties are the same in all directions (axial, lateral, and in between). Thus, the material contains an infinite number of planes of material property symmetry passing through a point. i.e., material properties are directionally independent. So, there are two independent elastic constants.

$$
\begin{Bmatrix} \sigma_{11} \\ \sigma_{22} \\ \sigma_{33} \\ \tau_{23} \\ \tau_{31} \\ \tau_{12} \end{Bmatrix} =
\begin{pmatrix}
C_{11} & C_{12} & C_{12} & 0 & 0 & 0 \\
C_{12} & C_{11} & C_{12} & 0 & 0 & 0 \\
C_{12} & C_{12} & C_{11} & 0 & 0 & 0 \\
0 & 0 & 0 & \dfrac{(C_{11}-C_{12})}{2} & 0 & 0 \\
0 & 0 & 0 & 0 & \dfrac{(C_{11}-C_{12})}{2} & 0 \\
0 & 0 & 0 & 0 & 0 & \dfrac{(C_{11}-C_{12})}{2}
\end{pmatrix}
\begin{Bmatrix} \varepsilon_{11} \\ \varepsilon_{22} \\ \varepsilon_{33} \\ \gamma_{23} \\ \gamma_{31} \\ \gamma_{12} \end{Bmatrix}
$$

Tensile normal stresses applied in any direction on an isotropic material cause only elongation in the direction of the applied stresses and contractions in the two transverse directions. It will not produce any shear strain in any form in the material. Similarly, shear stresses produce only corresponding shear strains not normal strains.

As the material properties are directionly independent, isotropic material has equal strength in all directions. As such, efficient structure (without wasting material structurally) is not possible by isotropic material (e.g., beam). In a beam the load is applied on the transverse direction and the beam will bend by extending and contracting in the lengthwise direction. In the lateral direction there is no load applied and the strength in the lateral direction is under utilized. Hence, isotropic material in beam structure is not efficient design, as the material is under utilized. On the other hand, unidirectional fibers which are aligned in the lengthwise direction will have high strength in the lengthwise direction and low strength in the other direction and thus, it becomes an efficient design. If the structure is taking complex loading system then, isotropic properties will be ideal and by arranging the fibers randomly quasi iso- tropic properties can be achieved.

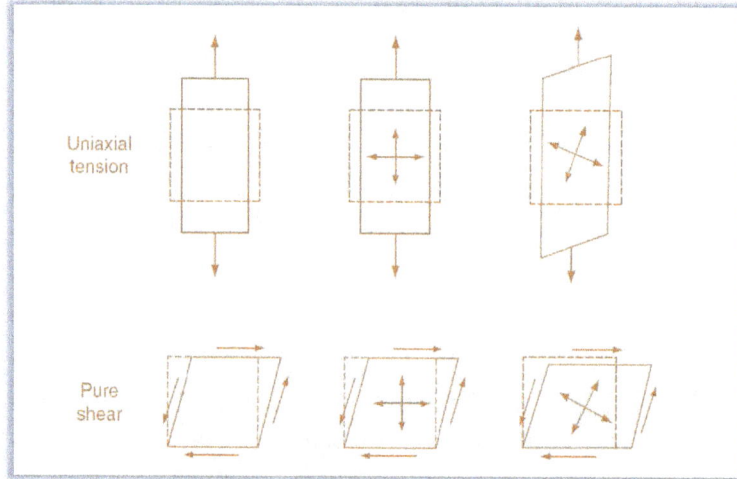

Differences in the deformations on uniaxial tension and pure shear

From figure, it is clear that isotropic and specially orthotropic materials behave in a similar way. But, the magnitude of deformation is direction dependent in the case of orthotropic whereas in the case of isotropic it is not. Anisotropic (and generally, orthotropic material) has coupling between normal and shear deformations.

To summarize the elastic constants of each type of material is given in Table.

Table: Elastic constants of different materials

Material	Three dimensional		Two dimensional	
	Number of non zero constants	Number of indepen- dent constants	Number of non zero constants	Number of indepen- dent constants
Anisotropic	36	21	9	6
Generally Orthotropic	36	9	9	4
Specially Orthotropic	12	9	5	4
Transversely Isotropic	12	5	5	4
Isotropic	12	2	5	2

Two-Dimensional Unidirectional Lamina

Consider, a thin plate of unidirectional composite material of uniform cross-section. If there are loads applied only along the edges and no out-of-plane loads, then, it can be considered to be a plane stress case.

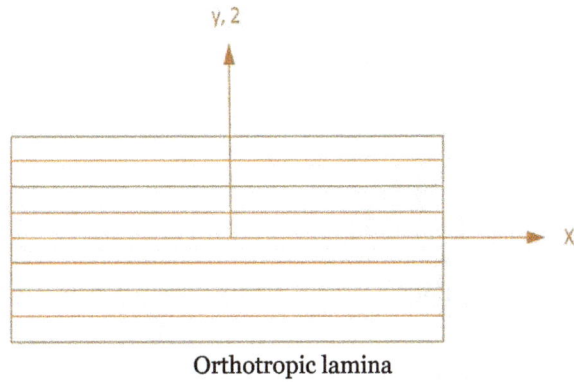

Orthotropic lamina

Since, the plate is very thin, the stresses normal to plane of loading i.e. σ_z, τ_{xz} and τ_{yz} can be assumed to vary insignificantly across the thickness. Thus, they can be assumed to be zero within the plate. i.e.

$$\sigma_z = 0, \tau_{xz} = 0, and\ \tau_{yz} = 0$$

This assumption reduces the three-dimensional stress–strain equations to two-dimensional stress–strain equations.

From generalized Hooke's law,

$$\varepsilon_z = \frac{1}{E}\Big[\sigma_z - v\big(\sigma_x + \sigma_y\big)\Big]$$

If σ_z = 0, then

$$\varepsilon_z = -\frac{v}{E}\big(\sigma_x + \sigma_y\big)$$

Let, the principal material directions be designated by the longitudinal direction L and the transverse direction T. Considering various loading conditions, the corresponding strains in the longitudinal and transverse directions are determined as follows.

Case (i) Only longitudinal load is applied

Therefore,

$$\sigma_L \neq 0, \sigma_T = \tau_{LT} = 0$$

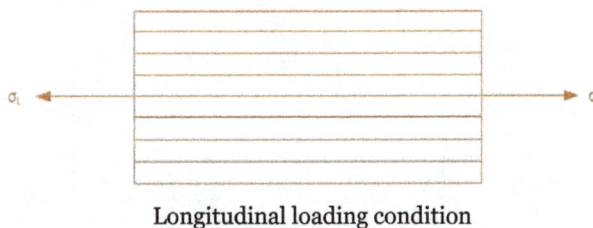

Longitudinal loading condition

The strains corresponding to this loading condition are,

$$\varepsilon_L = \frac{\sigma_L}{E_L}$$

$$\varepsilon_T = -v_{LT}\frac{\sigma_L}{E_L}$$

$$\gamma_{LT} = 0$$

Case (ii) Only transverse load is applied

Therefore,

$$\sigma_T \neq 0, \sigma_L = \tau_{LT} = 0$$

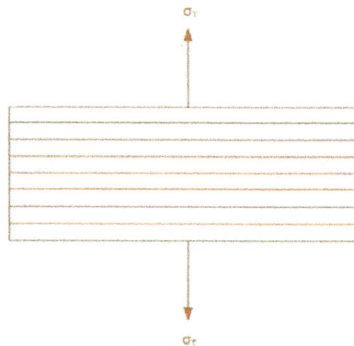

Transverse loading condition

The strains corresponding to this loading condition are,

$$\varepsilon_T = \frac{\sigma_T}{E_T}$$

$$\varepsilon_L = -v_{TL}\frac{\sigma_T}{E_T}$$

$$\gamma_{LT} = 0$$

Case (iii) Only shear load is applied

Therefore,

$$\tau_{LT} \neq 0, \quad \sigma_L = \sigma_T = 0$$

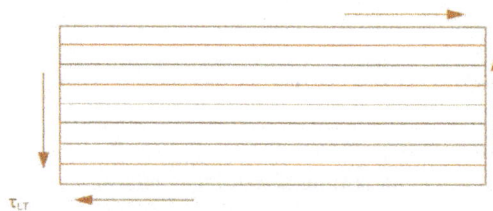

Shear loading condition

The strains corresponding to this loading condition are,

$$\gamma_{LT} = \frac{\tau_{LT}}{G_{LT}}$$

$$\varepsilon_L = \varepsilon_T = 0$$

Superimposing cases (i), (ii) and (iii)

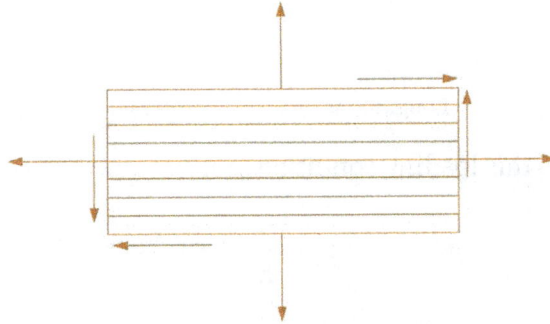

Lamina under loading condition

The strains corresponding to this loading condition are,

$$\varepsilon_L = \frac{\sigma_L}{E_L} - v_{TL} \frac{\sigma_T}{E_T}$$

$$\varepsilon_T = \frac{\sigma_T}{E_T} - v_{LT} \frac{\sigma_L}{E_L}$$

$$\gamma_{LT} = \frac{\tau_{LT}}{G_{LT}}$$

Anisotropic Layer (Generally Orthotropic)

When an orthotropic lamina is loaded in the direction other than its principal material axes, then, the behaviour of the lamina will be anisotropic. In the figure, the lamina will behave like anisotropic if load is applied along x or y directions.

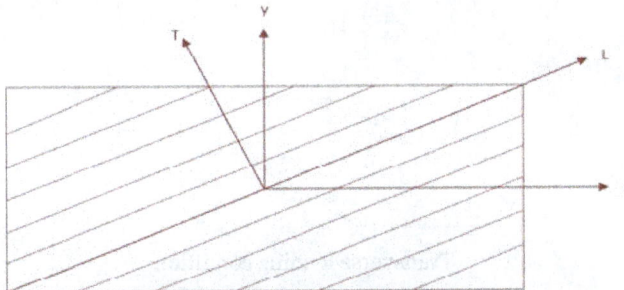

Generally Orthotropic lamina

Case (i) Only longitudinal load is applied (along x direction)

$$\sigma_x \neq 0,$$
$$\sigma_y = \tau_{xy} = 0$$

Therefore,

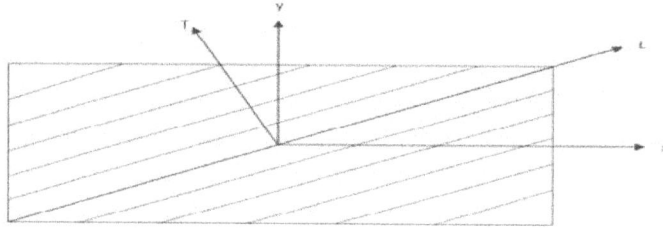

Longitudinal loading condition

The strains corresponding to this loading condition are,

$$\varepsilon_x = \frac{\sigma_x}{E_x}$$

$$\varepsilon_y = -v_{xy} \frac{\sigma_x}{E_x}$$

$$\gamma_{xy} = -m_x \frac{\sigma_x}{E_L}$$

where, m_x is the coefficient of mutual influence and

E_L is the modulus of composite along L direction

Case (ii) Only transverse load is applied

$$\sigma_y \neq 0$$
$$\sigma_x = \tau_{xy} = 0$$

Therefore,

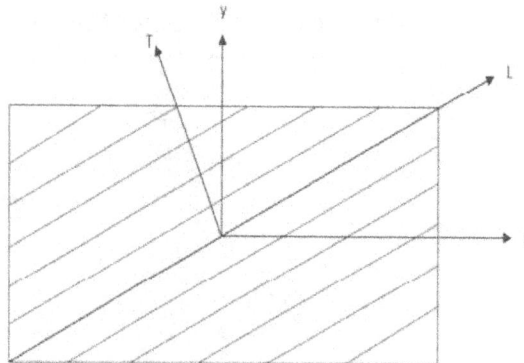

Transverse loading condition

The strains corresponding to this loading condition are,

$$\varepsilon_y = \frac{\sigma_y}{E_y}$$

$$\varepsilon_x = -v_{yx}\frac{\sigma_y}{E_y}$$

$$\gamma_{xy} = -m_y\frac{\sigma_y}{E_L}$$

Case (ii) Only shear load is applied

$$\tau_{xy} \neq 0$$

$$\sigma_x = \sigma_y = 0$$

Therefore,

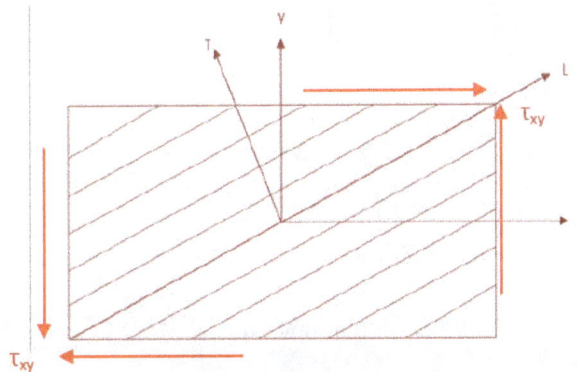

Shear loading condition

The strains corresponding to this loading condition are,

$$\varepsilon_x = -m_x\frac{\tau_{xy}}{E_L}$$

$$\varepsilon_y = -m_y\frac{\tau_{xy}}{E_L}$$

$$\gamma_{xy} = \frac{\tau_{xy}}{G_{xy}}$$

Case (iv) All loads are applied

Superimposing cases (i), (ii) and (iii)

$$\varepsilon_x = \frac{\sigma_x}{E_x} - v_{yx}\frac{\sigma_y}{E_y} - m_x\frac{\tau_{xy}}{E_L}$$

$$\varepsilon_y = \frac{\sigma_y}{E_y} - v_{xy}\frac{\sigma_x}{E_x} - m_y\frac{\tau_{xy}}{E_L}$$

$$\gamma_{xy} = \frac{\tau_{xy}}{G_{xy}} - m_x\frac{\sigma_x}{E_L} - m_y\frac{\sigma_y}{E_L}$$

Transformation of Engineering Constants

It is of interest to derive and know the explicit expressions for the usual engineering constants in the arbitrary axes in terms of those, along the principal material axes.

Consider an orthotropic lamina with its principal material axes (L and T) oriented at an angle θ with reference axes (x and y) as shown in Figure.

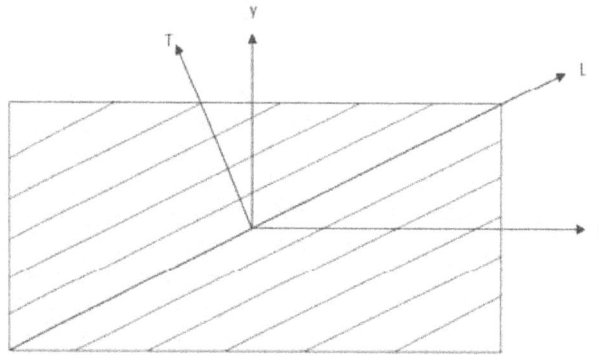

Orthotropic lamina

Let, the only nonzero stress acting on this lamina be σx. i.e.,

$$\sigma_x \neq 0$$
$$\sigma_y = \tau_{xy} = 0$$

The normal and shearing stresses along the L and T directions can be calculated by the stress-transformation law:

$$\sigma_L = \sigma_x \cos^2\theta + \sigma_y \sin 2\theta + 2\tau_{Xy}\sin\theta\cos\theta$$
$$\therefore \sigma_L = \sigma_x \cos^2\theta$$
$$\sigma_T = \sigma_x \sin^2\theta + \sigma_y \cos 2\theta - 2\tau_{Xy}\sin\theta\cos\theta$$

$$\therefore \sigma_T = \sigma_x \sin^2\theta$$
$$\tau_{LT} = -\sigma_x \sin\theta\cos\theta + \sigma_y \sin\theta\cos\theta + \tau_{xy}\left(\cos 2\theta - \sin 2\theta\right)$$
$$\therefore \tau_{LT} = -\sigma_x \sin\theta\cos\theta$$

The above relations are substituted in the strain relations derived previously for the orthotropic lamina in the L and T directions.

$$\varepsilon_L = \frac{\sigma_L}{E_L} - v_{TL} \frac{\sigma_T}{E_T}$$

$$\varepsilon_L = \frac{\sigma_x \cos^2 \theta}{E_L} - v_{TL} \frac{\sigma_x \sin^2 \theta}{E_T}$$

$$\varepsilon_T = \frac{\sigma_T}{E_T} - v_{LT} \frac{\sigma_L}{E_L}$$

$$\varepsilon_T = \frac{\sigma_x \sin^2 \theta}{E_T} - v_{LT} \frac{\sigma_x \cos^2 \theta}{E_L}$$

$$\gamma_{LT} = \frac{\tau_{LT}}{G_{LT}}$$

$$\gamma_{LT} = -\frac{\sigma_x \sin \theta \cos \theta}{G_{LT}}$$

The strains in the x and y directions can be obtained as by taking the inverse of the strain-transformation law, which can be written as:

$$\varepsilon_x = \varepsilon_L \cos^2 \theta + \varepsilon_T \sin^2 \theta - \gamma_{LT} \sin \theta \cos \theta$$

$$\varepsilon_y = \varepsilon_L \sin^2 \theta + \varepsilon_T \cos^2 \theta + \gamma_{LT} \sin \theta \cos \theta$$

$$\gamma_{xy} = 2(\varepsilon_L - \varepsilon_T) \sin \theta \cos \theta + \gamma_{LT} (\cos^2 \theta - \sin^2 \theta)$$

Substitution of the values of $\varepsilon_L, \varepsilon_T, and \, \gamma_{LT}$ the above equations become,

$$\varepsilon_x = \left(\frac{\sigma_x \cos^2 \theta}{E_L} - v_{TL} \frac{\sigma_x \sin^2 \theta}{E_T} \right) \cos^2 \theta + \left(\frac{\sigma_x \sin^2 \theta}{E_T} - v_{LT} \frac{\sigma_x \cos^2 \theta}{E_L} \right) \sin^2 \theta + \left(\frac{\sigma_x \sin \theta \cos \theta}{G_{LT}} \right) \sin \theta \cos \theta$$

$$= \sigma_x \left[\left(\frac{\cos^2 \theta}{E_L} - v_{LT} \frac{\sin^2 \theta}{E_T} \right) \cos^2 \theta + \left(\frac{\sin^2 \theta}{E_T} - v_{LT} \frac{\cos^2 \theta}{E_L} \right) \sin^2 \theta + \left(\frac{\sin \theta \cos \theta}{G_{LT}} \right) \sin \theta \cos \theta \right]$$

$$= \sigma_x \left[\frac{\cos^4 \theta}{E_L} - v_{TL} \frac{\sin^2 \theta \cos^2 \theta}{E_T} + \frac{\sin^4 \theta}{E_T} - v_{LT} \frac{\sin^2 \theta \cos^2 \theta}{E_L} + \frac{\sin^2 \theta \cos^2 \theta}{G_{LT}} \right]$$

$$= \sigma_x \left[\frac{\cos^4 \theta}{E_L} - v_{LT} \frac{\sin^2 \theta \cos^2 \theta}{E_L} + \frac{\sin^4 \theta}{E_T} - v_{LT} \frac{\sin^2 \theta \cos^2 \theta}{E_L} + \frac{\sin^2 \theta \cos^2 \theta}{G_{LT}} \right]$$

$$= \sigma_x \left[\frac{\cos^4 \theta}{E_L} + \frac{\sin^4 \theta}{E_T} - 2v_{LT} \frac{\sin^2 \theta \cos^2 \theta}{E_L} + \frac{\sin^2 \theta \cos^2 \theta}{G_{LT}} \right]$$

$$\varepsilon_x = \sigma_x \left[\frac{\cos^4 \theta}{E_L} + \frac{\sin^4 \theta}{E_T} + \frac{1}{4} \left(\frac{1}{G_{LT}} - \frac{2v_{LT}}{E_L} \right) \sin^2 2\theta \right]$$

$$\varepsilon_y = \varepsilon_L \sin^2 \theta + \varepsilon_T \cos^2 \theta + \gamma_{LT} \sin \theta \cos \theta$$

$$= \left(\frac{\sigma_x \cos^2 \theta}{E_L} - v_{TL} \frac{\sigma_x \sin^2 \theta}{E_T} \right) \sin^2 \theta + \left(\frac{\sigma_x \sin^2 \theta}{E_T} - v_{LT} \frac{\sigma_x \cos^2 \theta}{E_L} \right) \cos^2 \theta - \left(\frac{\sigma_x \sin \theta \cos \theta}{G_{LT}} \right) \sin \theta \cos \theta$$

$$= \sigma_x \left[\frac{\sin^2 \theta \cos^2 \theta}{E_L} - v_{TL} \frac{\sin^4 \theta}{E_T} + \frac{\sin^2 \theta \cos^2 \theta}{E_T} - v_{LT} \frac{\cos^4 \theta}{E_L} - \frac{\sin^2 \theta \cos^2 \theta}{G_{LT}} \right]$$

$$= \sigma_x \left[\left(\frac{1}{E_L} + \frac{1}{E_T} - \frac{1}{G_{LT}} \right) \sin^2 \theta \cos^2 \theta - \left(\sin^4 \theta + \cos^4 \theta \right) \frac{v_{LT}}{E_L} \right]$$

From trigonometry:

$$\left(\sin^2 \theta + \cos^2 \theta \right)^2 = \sin^4 \theta + \cos^4 \theta + 2 \sin^2 \theta \cos^2 \theta$$

$$\sin^4 \theta + \cos^4 \theta = 1 - 2 \sin^2 \theta \cos^2 \theta$$

$$= \sigma_x \left[\left(\frac{1}{E_L} + \frac{1}{E_T} - \frac{1}{G_{LT}} \right) \sin^2 \theta \cos^2 \theta - \left(1 - \sin^2 \theta \cos^2 \theta \right) \frac{v_{LT}}{E_L} \right]$$

$$= -\sigma_x \left[\frac{v_{LT}}{E_L} - \left(\frac{1}{E_L} + \frac{1}{E_T} + \frac{2v_{LT}}{E_L} - \frac{1}{G_{LT}} \right) \sin^2 \theta \cos^2 \theta \right]$$

$$\varepsilon_y = -\sigma_x \left[\frac{v_{LT}}{E_L} - \frac{1}{4} \left(\frac{1}{E_L} + \frac{1}{E_T} + \frac{2v_{LT}}{E_L} - \frac{1}{G_{LT}} \right) \sin^2 \theta \right]$$

$$\gamma_{xy} = 2 \left(\varepsilon_L - \varepsilon_T \right) \sin \theta \cos \theta + \gamma_{LT} \left(\cos^2 \theta - \sin^2 \theta \right)$$

$$\gamma_{xy} = 2 \left(\frac{\sigma_x \cos^2 \theta}{E_L} - v_{TL} \frac{\sigma_x \sin^2 \theta}{E_T} - \frac{\sigma_x \sin^2 \theta}{E_T} + v_{TL} \frac{\sigma_x \cos^2 \theta}{E_L} \right) \sin \theta \cos \theta - \frac{\sigma_x \sin \theta \cos \theta}{G_{LT}} \left(\cos^2 \theta - \sin^2 \theta \right)$$

$$= \sigma_x \left(\frac{2 \cos^3 \theta \sin \theta}{E_L} - v_{TL} \frac{2 \sin^3 \theta \cos \theta}{E_T} - \frac{2 \cos^3 \theta \cos \theta}{E_T} + v_{LT} \frac{2 \cos^3 \theta \sin \theta}{E_L} - \frac{\cos^3 \theta \sin \theta}{G_{LT}} + \frac{2 \sin^3 \theta \cos \theta}{G_{LT}} \right)$$

$$= \sigma_x \sin 2\theta \left(\frac{\cos^2 \theta}{E_L} - v_{TL} \frac{\sin^2 \theta}{E_T} - \frac{\sin^2 \theta}{E_T} + v_{LT} \frac{\cos^2 \theta}{E_L} - \frac{\cos^2 \theta}{2G_{LT}} + \frac{\sin^2 \theta}{2G_{LT}} \right)$$

$$\gamma_{xy} = \sigma_x \sin 2\theta \left[-\frac{v_{LT}}{E_L} - \frac{1}{E_T} + \frac{1}{2G_{LT}} + \cos^2 \theta \left(\frac{1}{E_L} + \frac{2v_{LT}}{E_L} + \frac{1}{E_T} - \frac{1}{G_{LT}} \right) \right]$$

To summarize,

$$\epsilon_x = \sigma_x \left[\frac{\cos^4 \theta}{E_L} + \frac{\sin^4 \theta}{E_T} + \frac{1}{4} \left(\frac{1}{G_{LT}} - \frac{2v_{LT}}{E_L} \right) \sin^2 2\theta \right]$$

$$\epsilon_y = -\sigma_x \left[\frac{v_{LT}}{E_L} - \frac{1}{4} \left(\frac{1}{E_L} + \frac{1}{E_T} + \frac{2v_{LT}}{E_L} - \frac{1}{G_{LT}} \right) \sin^2 2\theta \right]$$

$$\gamma_{xy} = \sigma_x \sin 2\theta \left[-\frac{v_{LT}}{E_L} - \frac{1}{E_T} + \frac{1}{2G_{LT}} + \cos^2 \theta \left(\frac{1}{E_L} + \frac{2v_{LT}}{E_L} + \frac{1}{E_T} - \frac{1}{G_{LT}} \right) \right]$$

Determinations of Elastic Constants

Once the strains are found, the elastic constants can be determined from the Hooke's law. The modulus of elasticity in the x-direction is given by:

$$E_x = \frac{\sigma_x}{\epsilon_x}$$

From the previous equation for ε_x, E_x can be calculated as :

$$\frac{1}{E_x} = \frac{\cos^4 \theta}{E_L} + \frac{\sin^4 \theta}{E_T} + \frac{1}{4} \left(\frac{1}{G_{LT}} - \frac{2v_{LT}}{E_L} \right) \sin^2 2\theta$$

The expression for E_y can be obtained by substituting $\theta + 90°$ for θ in above equation :

$$\frac{1}{E_y} = \frac{\sin^4 \theta}{E_L} + \frac{\cos^4 \theta}{E_T} + \frac{1}{4} \left(\frac{1}{G_{LT}} - \frac{2v_{LT}}{E_L} \right) \sin^2 2\theta$$

The Poisson ratio is defined as:

$$v_{xy} = -\frac{\varepsilon_y}{\varepsilon_x}$$

$$v_{xy} = -\varepsilon_y \left(\frac{E_x}{\sigma_x} \right)$$

$$v_{xy} = \left(\frac{E_x}{\sigma_x}\right)\sigma_x\left[\frac{v_{LT}}{E_L} - \frac{1}{4}\left(\frac{1}{E_L} + \frac{1}{E_T} + \frac{2v_{LT}}{E_L} - \frac{1}{G_{LT}}\right)\sin^2 2\theta\right]$$

$$\frac{v_{xy}}{E_x} = \frac{v_{LT}}{E_L} - \frac{1}{4}\left(\frac{1}{E_L} + \frac{1}{E_T} + \frac{2v_{LT}}{E_L} - \frac{1}{G_{LT}}\right)\sin^2 2\theta$$

Similarly,

$$\frac{v_{yx}}{E_y} = \frac{v_{TL}}{E_T} - \frac{1}{4}\left(\frac{1}{E_L} + \frac{1}{E_T} + \frac{2v_{LT}}{E_L} - \frac{1}{G_{LT}}\right)\sin^2 2\theta$$

When the normal stress σ_x is applied in a direction other than the L and T direction, it may induce a shearing strain given by the above equations. Therefore, a coefficient of mutual influence, m_x, may be defined that relates the shearing strain to the normal stress σ_x in the following manner:

$$\gamma_{xy} = -m_x\frac{\sigma_x}{E_L}$$

$$m_x = -\gamma_{xy}\frac{E_L}{\sigma_x}$$

Substituting the values σ_x and γ_{xy},

$$m_x = \sin 2\theta\left[v_{LT} + \frac{E_L}{E_T} - \frac{E_L}{2G_{LT}} - \cos^2\theta\left(1 + 2v_{LT} + \frac{E_L}{E_T} - \frac{E_L}{G_{LT}}\right)\right]$$

Similarly, the coefficient, m_y, which relates the shearing strain to normal stress σ_y is defined as :

$$\gamma_{xy} = -m_y\frac{\sigma_y}{E_L}$$

$$m_y = \sin 2\theta\left[v_{LT} + \frac{E_L}{E_T} - \frac{E_L}{2G_{LT}} - \sin^2\theta\left(1 + 2v_{LT} + \frac{E_L}{E_T} - \frac{E_L}{G_{LT}}\right)\right]$$

To obtain an expression for G_{xy}, assume that the only non-zero stress acting on the lamina is τ_{xy} (Pure shear case). The stresses along the principal material directions are given by :

$$\sigma_L = 2\tau_{xy}\sin\theta\cos\theta$$

$$\sigma_T = -2\tau_{xy}\sin\theta\cos\theta$$

$$\tau_{LT} = \tau_{xy}\left(\cos^2\theta - \sin^2\theta\right)$$

The corresponding strains are given by Hooke's law,

$$\varepsilon_L = \frac{\sigma_L}{E_L} - v_{TL}\frac{\sigma_T}{E_T}$$

$$\varepsilon_L = \left(\frac{1}{E_L} + \frac{v_{TL}}{E_T}\right)2\tau_{xy}\sin\theta\cos\theta$$

$$\varepsilon_T = \frac{\sigma_T}{E_T} - v_{LT}\frac{\sigma_L}{E_L}$$

$$\varepsilon_T = \left(\frac{1}{E_T} + \frac{v_{LT}}{E_L}\right)-2\tau_{xy}\sin\theta\cos\theta$$

$$\gamma_{LT} = \frac{\tau_{LT}}{G_{LT}}$$

$$\gamma_{LT} = \frac{\tau_{xy}}{G_{LT}}\left(\cos^2\theta - \sin^2\theta\right)$$

Substitution of the above equations gives the shearing strain γ_{xy},

$$\gamma_{xy} = 2\left(\varepsilon_L - \varepsilon_T\right)\sin\theta\cos\theta + \gamma_{LT}\left(\cos^2\theta - \sin^2\theta\right)$$

$$\gamma_{xy} = 2\left[\left(\frac{1}{E_L} + \frac{v_{TL}}{E_T}\right)2\tau_{xy}\sin\theta\cos\theta - \left(\frac{1}{E_T} + \frac{v_{LT}}{E_L}\right)-2\tau_{xy}\sin\theta\cos\theta\right]\sin\theta\cos\theta$$

$$+\frac{\tau_{xy}}{G_{LT}}\left(\cos^2\theta - \sin^2\theta\right)\left(\cos^2\theta - \sin^2\theta\right)$$

$$\gamma_{xy} = 4\tau_{xy}\left\{\left[\left(\frac{1}{E_L} + \frac{v_{TL}}{E_T}\right) - \left(\frac{1}{E_T} + \frac{v_{LT}}{E_L}\right)-1\right]\sin 2\theta\cos 2\theta + \frac{1}{G_{LT}}\left(\cos^2\theta - \sin^2\theta\right)2\right\}$$

$$\frac{1}{G_{xy}} = \frac{\gamma x_y}{\tau x_y} = \frac{1}{E_L} + \frac{2v_{LT}}{E_L} + \frac{1}{E_T} - \left(\frac{1}{E_L} + \frac{2v_{LT}}{E_L} + \frac{1}{E_T} - \frac{1}{G_{LT}}\right)\cos^2\theta$$

As the normal stresses do, the shearing stress τ_{xy} will also cause direct strains ε_x, and ε_y in the x and y directions, respectively, given by:

$$\varepsilon_x = -m_x\frac{\tau_{xy}}{E_L}$$

$$\varepsilon_y = -m_y \frac{\tau_{xy}}{E_L}$$

It will be relevant at this point to note that the stress-strain relations for an orthotropic lamina referred to arbitrary axes can be written in terms of engineering constants as:

$$\varepsilon_x = \frac{\sigma_x}{E_x} - v_{yx} \frac{\sigma_y}{E_y} - m_x \frac{\tau_{xy}}{E_L}$$

$$\varepsilon_y = \frac{\sigma_y}{E_y} - v_{xy} \frac{\sigma_x}{E_x} - m_y \frac{\tau_{xy}}{E_L}$$

$$\gamma_{xy} = \frac{\tau_{xy}}{G_{xy}} - m_x \frac{\sigma_x}{E_L} - m_y \frac{\sigma_y}{E_L}$$

The elastic constants E_x, E_y, v_{xy}, v_{yx}, m_x, m_y, and G_{xy} are determined from the equations given above.

Specially Orthotropic Material Under Plane Stress

In plane stress condition, the stresses normal to the plane under consideration are assumed to be zero. i.e., if the 1-2 (xy) plane is assumed to be the loading plane, then the normal stresses to 1-2 plane

$$\sigma_3 = \tau_{23} = \tau_{13} = 0$$

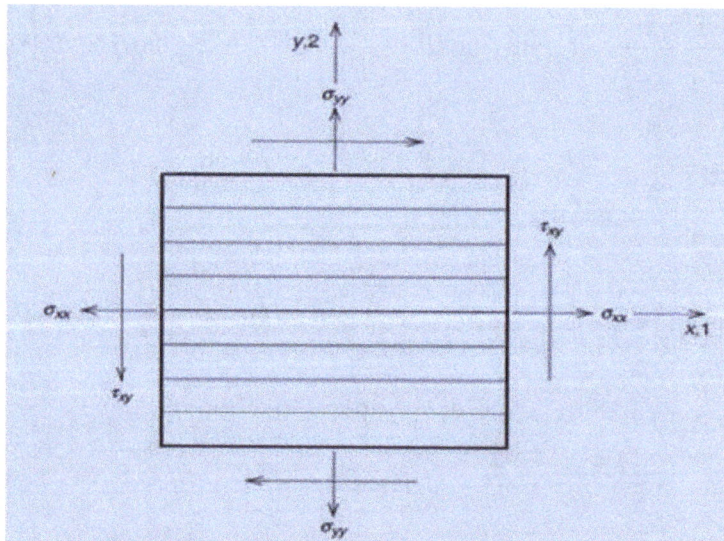

Orthotropic lamina

More over, the number of independent elastic constants for orthotropic material in the plane stress condition are reduced to 4 from 9 in the three-dimensional case. Thus, the stress-strain relation is given by

$$
\begin{Bmatrix} \sigma_{11} \\ \sigma_{22} \\ \sigma_{33}=0 \\ \tau_{23}=0 \\ \tau_{31}=0 \\ \tau_{12} \end{Bmatrix}
\begin{pmatrix}
C_{11} & C_{12} & C_{13} & 0 & 0 & 0 \\
C_{12} & C_{22} & C_{23} & 0 & 0 & 0 \\
C_{13} & C_{23} & C_{33} & 0 & 0 & 0 \\
0 & 0 & 0 & C_{44} & 0 & 0 \\
0 & 0 & 0 & 0 & C_{55} & 0 \\
0 & 0 & 0 & 0 & 0 & C_{66}
\end{pmatrix}
\begin{Bmatrix} \epsilon_{11} \\ \epsilon_{22} \\ \epsilon_{33} \\ \gamma_{23} \\ \gamma_{31} \\ \gamma_{12} \end{Bmatrix}
$$

$$
\sigma_{11} = C_{11}\,\varepsilon_{11} + C_{12}\,\varepsilon_{22} + C_{13}\,\varepsilon_{33}
$$

$$
\sigma_{22} = C_{12}\,\varepsilon_{11} + C_{22}\,\varepsilon_{22} + C_{23}\,\varepsilon_{33}
$$

$$
0 \quad = C_{13}\,\varepsilon_{11} + C_{32}\,\varepsilon_{22} + C_{33}\,\varepsilon_{33}
$$

$$
\gamma_{23} = 0
$$

$$
\gamma_{13} = 0
$$

$$
\tau_{12} = C_{66}\,\gamma_{12}
$$

after eliminating ε_{33}, the equations may be written as

$$
\sigma_{11}\ or\ \sigma_1 = \left(c_{11} - \frac{C_{132}}{C_{33}} \right)\varepsilon_1 + \left(c_{12} - \frac{C_{13}C_{23}}{C_{33}} \right)\varepsilon_2
$$

$$
\sigma_{22}\ or\ \sigma_2 = \left(c_{12} - \frac{c_{13}c_{23}}{c_{33}} \right)\varepsilon_1 + \left(c_{22} - \frac{C_{232}}{C_{33}} \right)\varepsilon_2
$$

$$
\tau_{12} = C_{66}\,\gamma_{12}
$$

or,

$$
\begin{Bmatrix} \sigma_1 \\ \sigma_2 \\ \tau_{12} \end{Bmatrix} =
\begin{bmatrix}
Q_{11} & Q_{12} & 0 \\
Q_{12} & Q_{22} & 0 \\
0 & 0 & Q_{66}
\end{bmatrix}
\begin{Bmatrix} \varepsilon_1 \\ \varepsilon_2 \\ \gamma_{12} \end{Bmatrix}
$$

Where,

$$
Q_{11} = \left(c_{11} - \frac{C_{132}}{C_{33}} \right)
$$

$$Q_{22} = \left(c_{22} - \frac{C_{232}}{C_{33}} \right)$$

$$Q_{12} = \left(c_{12} - \frac{C_{13}C_{23}}{C_{33}} \right)$$

$$Q_{66} = G_{12}$$

for specially orthotropic composite materials, the stiffness coefficients may be related with engineering constants as follows:

$$Q_{11} = \frac{E_L}{1 - v_{LT} v_{TL}}$$

$$Q_{22} = \frac{E_T}{1 - v_{LT} v_{TL}}$$

$$Q_{12} = \frac{v_{LT} E_T}{1 - v_{LT} v_{TL}} = \frac{v_{TL} E_L}{1 - v_{LT} v_{TL}}$$

$$Q_{66} = G_{LT}$$

The strain-stress relationship is given by,

$$\begin{Bmatrix} \varepsilon_1 \\ \varepsilon_2 \\ \gamma_{12} \end{Bmatrix} = \begin{bmatrix} S_{11} & S_{12} & 0 \\ S_{12} & S_{22} & 0 \\ 0 & 0 & S_{66} \end{bmatrix} \begin{Bmatrix} \sigma_1 \\ \sigma_2 \\ \tau_{12} \end{Bmatrix}$$

Where

$$S_{11} = \frac{1}{E_{11}}$$

$$S_{22} = \frac{1}{E_{22}}$$

$$S_{12} = -\frac{v_{12}}{E_{11}} = -\frac{v_{21}}{E_{22}}$$

$$S_{66} = \frac{1}{G_{12}}$$

Stress-strain Relations for thin Isotropic Lamina

Stresses in an isotropic lamina under a plane stress condition is given by

$$
\begin{Bmatrix} \sigma_1 \\ \sigma_2 \\ \tau_{12} \end{Bmatrix} = \begin{bmatrix} Q_{11} & Q_{12} & 0 \\ Q_{12} & Q_{22} & 0 \\ 0 & 0 & Q_{66} \end{bmatrix} \begin{Bmatrix} \varepsilon_1 \\ \varepsilon_2 \\ \gamma_{12} \end{Bmatrix}
$$

Where

$$
Q_{11} = Q_{22} = \frac{E}{1-v^2}
$$

$$
Q_{12} = \frac{vE}{1-v^2}
$$

$$
Q_{66} = G = \frac{E}{2(1+v)}
$$

and the strain-stress relation is given by

$$
\begin{Bmatrix} \varepsilon_1 \\ \varepsilon_2 \\ \gamma_{12} \end{Bmatrix} = \begin{bmatrix} S_{11} & S_{12} & 0 \\ S_{12} & S_{22} & 0 \\ 0 & 0 & S_{66} \end{bmatrix} \begin{Bmatrix} \sigma_1 \\ \sigma_2 \\ \tau_{12} \end{Bmatrix}
$$

Where,

$$
S_{11} = S_{22} = \frac{1}{E}
$$

$$
S_{12} = -\frac{v}{E}
$$

$$
S_{66} = \frac{1}{G}
$$

Stress-strain Relations for Lamina with Arbitrary Orientation

Consider an orthotropic lamina with its principal material axes oriented at an angle θ with the reference coordinate axes as shown in figure. Stresses and strains can be easily transformed from one set of axes to another.

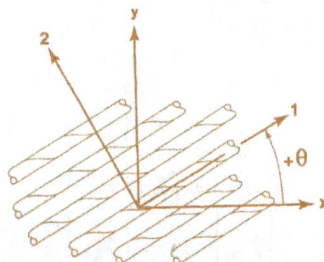

Lamina with arbitrary orientation

From elementary mechanics of materials the transformation equations for expressing stresses in a 1-2 coordinate system in terms of stresses in a x-y coordinate system,

$$\begin{Bmatrix} \sigma_1 \\ \sigma_2 \\ \tau_{12} \end{Bmatrix} = [T] \begin{Bmatrix} \sigma_x \\ \sigma_y \\ \tau_{xy} \end{Bmatrix}$$

$$[T] = \begin{bmatrix} \cos^2 \theta & \sin^2 \theta & 2\sin\theta\cos\theta \\ \sin^2 \theta & \cos^2 \theta & -2\sin\theta\cos\theta \\ -\sin\theta\cos\theta & \sin\theta\cos\theta & \cos^2\theta - \sin^2\theta \end{bmatrix}$$

The transformations are commonly written as

$$\begin{Bmatrix} \sigma_x \\ \sigma_y \\ \tau_{xy} \end{Bmatrix} = [T]^{-1} \begin{Bmatrix} \sigma_1 \\ \sigma_2 \\ \tau_{12} \end{Bmatrix}$$

$$\begin{Bmatrix} \varepsilon_x \\ \varepsilon_y \\ \dfrac{\gamma_{xy}}{2} \end{Bmatrix} = [T]^{-1} \begin{Bmatrix} \varepsilon_1 \\ \varepsilon_2 \\ \dfrac{\gamma_{12}}{2} \end{Bmatrix}$$

where, the transformation matrix $[T]^{-1}$ is given by

$$[T]^{-1} = \begin{bmatrix} \cos^2 \theta & \sin^2 \theta & -2\sin\theta\cos\theta \\ \sin^2 \theta & \cos^2 \theta & 2\sin\theta\cos\theta \\ \sin\theta\cos\theta & -\sin\theta\cos\theta & \cos^2\theta - \sin^2\theta \end{bmatrix}$$

$$\begin{Bmatrix} \varepsilon_1 \\ \varepsilon_2 \\ \gamma_{12} \end{Bmatrix} = [R] \begin{Bmatrix} \varepsilon_1 \\ \varepsilon_2 \\ \dfrac{\gamma_{12}}{2} \end{Bmatrix}$$

$$\begin{Bmatrix} \varepsilon_x \\ \varepsilon_y \\ \gamma_{xy} \end{Bmatrix} = [R] \begin{Bmatrix} \varepsilon_x \\ \varepsilon_y \\ \dfrac{\gamma_{xy}}{2} \end{Bmatrix}$$

where, $[R]$ is the Reuter matrix and is defined as

$$[R] = \begin{bmatrix} 1 & 0 & 0 \\ 0 & 1 & 0 \\ 0 & 0 & 2 \end{bmatrix}$$

For a specially orthotropic lamina whose principal material axes are aligned with the natural body axes,

$$\begin{Bmatrix} \sigma_1 \\ \sigma_2 \\ \tau_{12} \end{Bmatrix} = \begin{bmatrix} Q_{11} & Q_{12} & 0 \\ Q_{12} & Q_{22} & 0 \\ 0 & 0 & Q_{66} \end{bmatrix} \begin{Bmatrix} \varepsilon_1 \\ \varepsilon_2 \\ \gamma_{12} \end{Bmatrix} = [Q] \begin{Bmatrix} \varepsilon_1 \\ \varepsilon_2 \\ \gamma_{12} \end{Bmatrix}$$

$$\begin{Bmatrix} \sigma_x \\ \sigma_y \\ \tau_{xy} \end{Bmatrix} = [T]^{-1} \begin{Bmatrix} \sigma_1 \\ \sigma_2 \\ \tau_{12} \end{Bmatrix} = [T]^{-1} [Q] \begin{Bmatrix} \varepsilon_1 \\ \varepsilon_2 \\ \gamma_{12} \end{Bmatrix}$$

$$\begin{Bmatrix} \sigma_x \\ \sigma_y \\ \tau_{xy} \end{Bmatrix} = [T]^{-1} [Q][R] \begin{Bmatrix} \varepsilon_1 \\ \varepsilon_2 \\ \dfrac{\gamma_{12}}{2} \end{Bmatrix}$$

$$\begin{Bmatrix} \sigma_x \\ \sigma_y \\ \tau_{xy} \end{Bmatrix} = [T]^{-1} [Q][R][T] \begin{Bmatrix} \varepsilon_x \\ \varepsilon_y \\ \dfrac{\gamma_{xy}}{2} \end{Bmatrix}$$

$$\begin{Bmatrix} \sigma_x \\ \sigma_y \\ \tau_{xy} \end{Bmatrix} = [T]^{-1} [Q][R][T][R]^{-1} \begin{Bmatrix} \varepsilon_x \\ \varepsilon_y \\ \gamma_{xy} \end{Bmatrix}$$

However, $[R][T][R]^{-1} = [T]^{-T}$

$$\begin{Bmatrix} \sigma_y \\ xy \end{Bmatrix} \quad [T]^{-} [Q][T]^{-} \begin{Bmatrix} \varepsilon_y \\ xy \end{Bmatrix}$$

$$\begin{Bmatrix} \sigma_x \\ \sigma_y \\ \tau_{xy} \end{Bmatrix} = \begin{bmatrix} \bar{Q} \end{bmatrix} \begin{Bmatrix} \varepsilon_x \\ \varepsilon_y \\ \gamma_{xy} \end{Bmatrix}$$

where, $\begin{bmatrix} Q \end{bmatrix}$ is the transformed reduced stiffness matrix.

$$\begin{bmatrix} \bar{Q} \end{bmatrix} = \begin{bmatrix} (T) \end{bmatrix}^{\dagger}(-1)\begin{bmatrix} Q \end{bmatrix}\begin{bmatrix} (T) \end{bmatrix}^{\dagger}(-T)$$

Thus, the stress-strain relations in x-y coordinates are

$$\begin{Bmatrix} \sigma_x \\ \sigma_y \\ \tau_{xy} \end{Bmatrix} = \begin{bmatrix} \bar{Q}_{11} & \bar{Q}_{12} & \bar{Q}_{16} \\ \bar{Q}_{12} & \bar{Q}_{22} & \bar{Q}_{26} \\ \bar{Q}_{16} & \bar{Q}_{26} & \bar{Q}_{66} \end{bmatrix} \begin{Bmatrix} \varepsilon_x \\ \varepsilon_y \\ \gamma_{xy} \end{Bmatrix}$$

Where,

$$\bar{Q}_{11} = Q_{11}\cos^4\theta + Q_{22}\sin^4\theta + 2(Q_{12} + 2Q_{66})\sin^2\theta\cos^2\theta$$
$$\bar{Q}_{22} = Q_{11}\sin^4\theta + Q_{22}\cos^4\theta + 2(Q_{12} + 2Q_{66})\sin^2\theta\cos^2\theta$$
$$\bar{Q}_{12} = (Q_{11} + Q_{22} - 4Q_{66})\sin^2\theta\cos^2\theta + Q_{12}\left(\sin^4\theta\cos^4\theta\right)$$
$$\bar{Q}_{16} = (Q_{11} + Q_{12} - 2Q_{66})\sin\theta\cos^3\theta - (Q_{22} - Q_{12} - 2Q_{66})\sin^3\theta\cos\theta$$
$$\bar{Q}_{26} = (Q_{11} + Q_{12} - 2Q_{66})\sin^3\theta + \cos\theta - (Q_{22} - Q_{12} - 2Q_{66})\sin\theta\cos^3\theta$$
$$\bar{Q}_{66} = (Q_{11} + Q_{22} - 2Q_{12} - 2Q_{66})\sin^2\theta\cos^2\theta + Q_{66}\left(\sin^4\theta\cos^4\theta\right)$$

If $\theta = 0°$,

$$\bar{Q}_{11} = Q_{11}$$
$$\bar{Q}_{22} = Q_{22}$$
$$\bar{Q}_{11} = Q_{12}$$
$$\bar{Q}_{16} = \bar{Q}_{26} = 0$$
$$\bar{Q}_{66} = Q_{66}$$

If $\theta = 90°$,

$$\bar{Q}_{11} = Q_{22}$$
$$\bar{Q}_{22} = Q_{11}$$
$$\bar{Q}_{12} = Q_{12}$$
$$\bar{Q}_{16} = \bar{Q}_{26} = 0$$
$$\bar{Q}_{66} = Q_{66}$$

Similarly, the strain-stress relation along the principal axes for a specially orthotropic lamina can be written as:

$$\begin{Bmatrix} \varepsilon_1 \\ \varepsilon_2 \\ \gamma_{12} \end{Bmatrix} = \begin{bmatrix} S_{11} & S_{12} & 0 \\ S_{12} & S_{22} & 0 \\ 0 & 0 & S_{66} \end{bmatrix} \begin{Bmatrix} \sigma_1 \\ \sigma_2 \\ \tau_{12} \end{Bmatrix}$$

Therefore, the strain-stress relation along any arbitrary direction is given by

$$\begin{Bmatrix} \varepsilon_x \\ \varepsilon_y \\ \gamma_{xy} \end{Bmatrix} = \begin{bmatrix} \hat{S} \end{bmatrix} \begin{Bmatrix} \sigma_x \\ \sigma_y \\ \tau_{xy} \end{Bmatrix}$$

where, Ŝ is the transformed reduced compliance matrix

$$\hat{S} = [T]^T [S][T] \text{ and}$$

$$[T]^T = [R][T]^{-1}[R]^{-1}$$

Thus, the strain-stress relations in x-y coordinates are

$$\begin{Bmatrix} \varepsilon_x \\ \varepsilon_y \\ \gamma_{xy} \end{Bmatrix} = \begin{bmatrix} \hat{S}_{11} & \hat{S}_{12} & \hat{S}_{16} \\ \hat{S}_{12} & \hat{S}_{22} & \hat{S}_{26} \\ \hat{S}_{16} & \hat{S}_{26} & \hat{S}_{66} \end{bmatrix} \begin{Bmatrix} \sigma_x \\ \sigma_y \\ \tau_{xy} \end{Bmatrix}$$

where,

$$\bar{S}_{11} = S_{11} \cos^4 \theta + S_{22} \sin^4 \theta + 2(S_{12} + 2S_{66}) \sin^2 \theta \cos^2 \theta$$
$$\bar{S}_{22} = S_{11} \sin^4 \theta + S_{22} \cos^4 \theta + 2(S_{12} + 2S_{66}) \sin^2 \theta \cos^2 \theta$$
$$\bar{S}_{12} = (S_{11} + S_{22} - S_{66}) \sin^2 \theta \cos^2 \theta + S_{12} (\sin^4 \theta + \cos^4 \theta)$$
$$\bar{S}_{16} = (2S_{11} - 2S_{22} - S_{66}) \sin \theta \cos^2 \theta - (2S_{22} - 2S_{12} - S_{66}) \sin^2 \theta \cos \theta$$
$$\bar{S}_{26} = (2S_{11} - 2S_{12} - SQ_{66}) \sin^2 \theta \cos \theta - (2S_{22} - 2S_{12} - S_{66}) \sin \theta \cos^2 \theta$$
$$\bar{S}_{66} = 2(2S_{11} + 2S_{22} - 4S_{12} - S_{66}) \sin^2 \theta \cos^2 \theta - S_{66} (\sin^4 \theta + \cos^4 \theta)$$

If $\theta = 90°$,

$$\hat{S}_{11} = S_{11}; \ \hat{S}_{22} = S_{22}; \ \hat{S}_{12} = S_{12}; \hat{S}_{16} = \hat{S}_{26} = 0; \ v_{66} = S_{66}$$

If $\theta = 90°$,

$$\hat{S}_{11} = S_{22};\ \hat{S}_{22} = S_{11};\ \hat{S}_{12} = S_{12};\ \hat{S}_{16} = \hat{S}_{26} = 0;\ \hat{S}_{66} = S_{66}$$

Mechanical Properties of Composites

The mechanical properties of composite materials are determined by conducting standard tests framed by the American Society for Testing and Materials (ASTM). The followings are the list of ASTM standards, which are used to find the mechanical properties of composite materials.

Tensile test - ASTM D3039

Compression test - ASTM D3410

Flexural test - ASTM D790

±45 Shear test - ASTM D3518

In-plane Shear test - ASTM D4255

Inter-laminar shear strength - ASTM D2344

Tensile Strength Test

The tensile properties of composite laminate are determined in accordance with ASTM D3039. The tensile specimen is straight-sided and has the cross-section as shown in the figure.

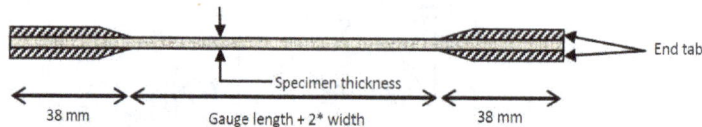

Tensile test specimen

The tensile specimen is held in the testing machine by wedge action grips. Longitudinal and transverse strains are measured using strain gauges. Longitudinal tensile modulus E_{11} and the major Poisson's ratio v_{12} are determined from the tension test data of 0° unidirectional laminates. The transverse modulus E_{22} and the minor Poisson's ratio v_{21} are determined from the tension test data of 900 unidirectional laminates.

Tensile test apparatus

For an off-axis unidirectional specimen ($0° < \theta < 90°$), a tensile load creates both extension and shear deformations (since A_{16} and A_{26} are not equal to zero). Hence, the experimentally determined modulus of an off-axis specimen is corrected to obtain its true modulus using the equation:

$$E_{true} = (1-\eta) E_{experimental}$$

$$\text{where, } \eta = 3 * S_{16}^2 / \left(S_{11}^2 \left[3*(S_{66}/S_{11}) + 2*(L/w)^2 \right] \right)$$

L - the specimen length between grips

w - the specimen width

S_{ij} - the elements in the compliance matrix

Flexural Strength Test

Flexural strength and modulus are determined by ASTM test method D790. In this test, a rectangular cross section of composite beam speciment is loaded in a three-point bending mode.

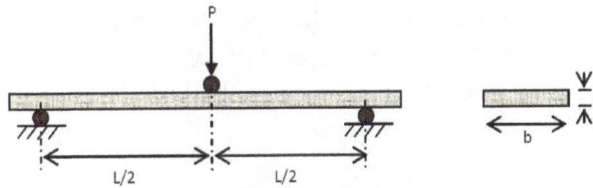

Flexural test setup

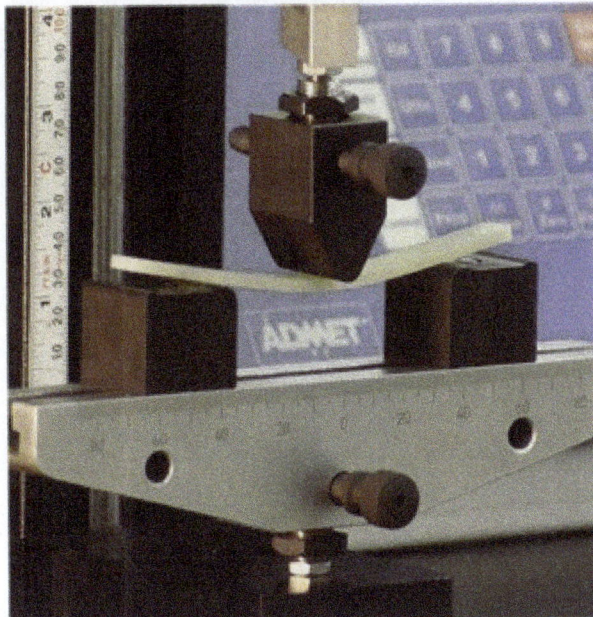

Flexural test apparatus - Three point loading

The maximum fiber stress at failure on the tension side of a flexural specimen is considered the flexural strength of the material. Therefore, the flexural strength is given by:

$$\sigma = \frac{M}{I} y$$

$$\sigma = \frac{\left(\dfrac{P}{2}\right)\left(\dfrac{l}{2}\right)\left(\dfrac{d}{2}\right)}{\dfrac{bd^2}{12}}$$

$$\sigma = \frac{3Pl}{2bd^2}$$

Inter-laminar Shear Strength (ILSS)

Inter-laminar shear strength of composite is the shear strength parallel to the plane of laminate. It is determined in accordance with ASTM D2344 using a short-beam shear test. It is applicable to all types of parallel fiber reinforced plastics and composites.

The thickness and width of the test specimen are measured before conditioning. The specimen is placed on a horizontal shear test fixture so that the fibers are parallel to the loading nose. The loading nose is then used to flex the specimen at a speed of .05 inches per minute until breakage. The force is then recorded. To determine shear strength, calculations are performed as given below.

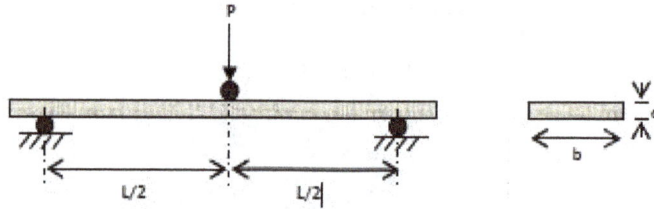

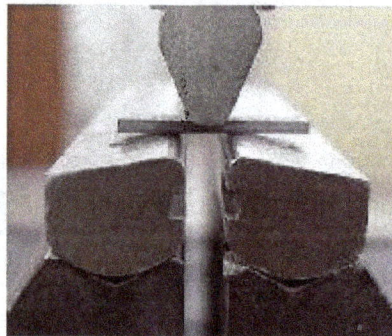

Short beam shear test

$$shear\ strength, \tau = \frac{V}{I} \int y\,ds$$

$$\tau = \frac{V}{I} \left(\frac{y^2}{2}\right)_0^{\frac{d}{2}}$$

$$\tau = \frac{Pd^2}{16\left(\dfrac{bd^2}{12}\right)}$$

$$\tau = \frac{3P}{4bd}$$

In-Plane Shear Strength (IPSS)

The shear modulus G_{12} and the ultimate shear strength τ_{12U} of unidirectional fiber-reinforced composites may be determined by any one of the following test procedures.

(i) $\pm 45°$ shear test (ASTM D3518)

(ii) Iosipescu or V-notched shear test (ASTM D5379)

(iii) Two-rail or three-rail shear test (ASTM D4255)

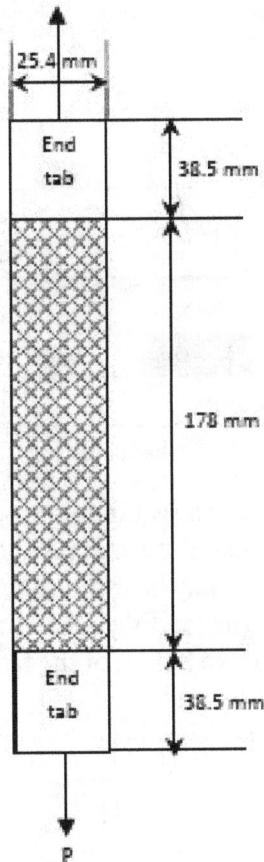

$\pm 45°$ shear test configuration

In $\pm 45°$ shear test, test specimens are placed in the grips of a universal tester at a specified grip separation and pulled until failure. Optional tabs can be bonded to the ends of the specimen to prevent gripping damage. The expressions to find the shear modulus and the shear strength is given by the expressions

$$G_{12} = \frac{\sigma_{xx}}{2\left(\varepsilon_{xx} - \varepsilon_{yy}\right)}$$

The shear strength is given by the expression

$$S_{12} = \frac{P_{max}}{2bh}$$

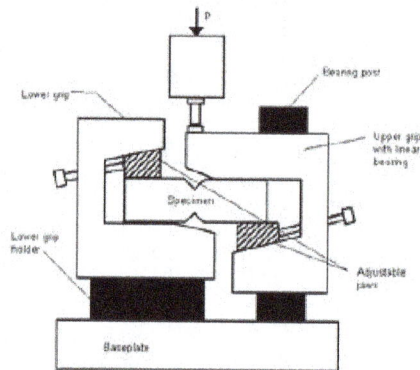

V-notched shear test

The Iosipescu or V-notched shear test uses a rectangular beam with symmetrical centrally located V-notches. The beam is loaded by a special fixture applying a shear loading at the V notch. Either in-plane or out-of-plane shear properties may be evaluated, depending upon the orientation of the material co-ordinate system relative to the loading axis. The notched specimen is loaded by introducing a relative displacement between two halves of the test fixture. The shear stress is calculated as

$$\tau = \frac{P}{wh}$$

where P - applied load

      w - distance between the notches

      h - specimen thickness

Two-rail shear test

In two-rail shear test, two pairs of steel rails are fastened along the long edges of a 76.2 mm wide and 152.4 mm long rectangular specimen, usually by three bolts on each side. At the other two edges, the specimen remains free. A tensile load is applied to the rails. That will induce an in-plane shear load on the tested laminate. The shear strength is calculated by using the expression

$$\tau = \frac{P}{Lh}$$

where, L - the specimen length

      h - the specimen thickness

Concepts of Solid Mechanics

Solid mechanics focuses on the behaviour of solid and composite materials. It is very important in fields like civil, nuclear and mechanical engineering. Euler-Bernoulli Bearn equation is one of the application of solid mechanics. The section strategically encompasses and incorporates the major components and key concepts of solid mechanics, providing a complete understanding.

Solid Mechanics

Solid mechanics is the branch of continuum mechanics that studies the behavior of solid materials, especially their motion and deformation under the action of forces, temperature changes, phase changes, and other external or internal agents.

Solid mechanics is fundamental for civil, aerospace, nuclear, and mechanical engineering, for geology, and for many branches of physics such as materials science. It has specific applications in many other areas, such as understanding the anatomy of living beings, and the design of dental prostheses and surgical implants. One of the most common practical applications of solid mechanics is the Euler-Bernoulli beam equation. Solid mechanics extensively uses tensors to describe stresses, strains, and the relationship between them.

Relationship to Continuum Mechanics

As shown in the following table, solid mechanics inhabits a central place within continuum mechanics. The field of rheology presents an overlap between solid and fluid mechanics.

Continuum mechanics The study of the physics of continuous materials	Solid mechanics The study of the physics of continuous materials with a defined rest shape.	Elasticity Describes materials that return to their rest shape after applied stresses are removed.	
		Plasticity Describes materials that permanently deform after a sufficient applied stress.	Rheology The study of materials with both solid and fluid characteristics.
	Fluid mechanics The study of the physics of continuous materials which deform when subjected to a force.	Non-Newtonian fluids do not undergo strain rates proportional to the applied shear stress.	
		Newtonian fluids undergo strain rates proportional to the applied shear stress.	

Response Models

A material has a rest shape and its shape departs away from the rest shape due to stress. The amount of departure from rest shape is called deformation, the proportion of deformation to original size is called strain. If the applied stress is sufficiently low (or the imposed strain is small enough), almost all solid materials behave in such a way that the strain is directly proportional to the stress; the coefficient of the proportion is called the modulus of elasticity. This region of deformation is known as the linearly elastic region.

It is most common for analysts in solid mechanics to use linear material models, due to ease of computation. However, real materials often exhibit non-linear behavior. As new materials are used and old ones are pushed to their limits, non-linear material models are becoming more common.

There are four basic models that describe how a solid responds to an applied stress:

1. Elastically – When an applied stress is removed, the material returns to its undeformed state. Linearly elastic materials, those that deform proportionally to the applied load, can be described by the linear elasticity equations such as Hooke's law.

2. Viscoelastically – These are materials that behave elastically, but also have damping: when the stress is applied and removed, work has to be done against the damping effects and is converted in heat within the material resulting in a hysteresis loop in the stress–strain curve. This implies that the material response has time-dependence.

3. Plastically – Materials that behave elastically generally do so when the applied stress is less than a yield value. When the stress is greater than the yield stress, the material behaves plastically and does not return to its previous state. That is, deformation that occurs after yield is permanent.

4. Thermoelastically - There is coupling of mechanical with thermal responses. In general, thermoelasticity is concerned with elastic solids under conditions that are neither isothermal nor adiabatic. The simplest theory involves the Fourier's law of heat conduction, as opposed to advanced theories with physically more realistic models.

Timeline

Galileo Galilei published the book "Two New Sciences" in which he examined the failure of simple structures

- 1452–1519 Leonardo da Vinci made many contributions

- 1638: Galileo Galilei published the book "Two New Sciences" in which he examined the failure of simple structures

- 1660: Hooke's law by Robert Hooke

- 1687: Isaac Newton published "Philosophiae Naturalis Principia Mathematica" which contains Newton's laws of motion

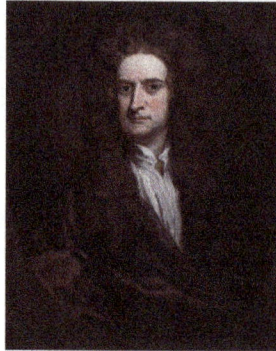

Isaac Newton published "Philosophiae Naturalis Principia Mathematica" which contains the Newton's laws of motion

- 1750: Euler–Bernoulli beam equation

- 1700–1782: Daniel Bernoulli introduced the principle of virtual work

- 1707–1783: Leonhard Euler developed the theory of buckling of columns

Leonhard Euler developed the theory of buckling of columns

- 1826: Claude-Louis Navier published a treatise on the elastic behaviors of structures

- 1873: Carlo Alberto Castigliano presented his dissertation "Intorno ai sistemi elastici", which contains his theorem for computing displacement as partial derivative of the strain energy. This theorem includes the method of *least work* as a special case

- 1874: Otto Mohr formalized the idea of a statically indeterminate structure.

- 1922: Timoshenko corrects the Euler-Bernoulli beam equation

- 1936: Hardy Cross' publication of the moment distribution method, an important innovation in the design of continuous frames.

- 1941: Alexander Hrennikoff solved the discretization of plane elasticity problems using a lattice framework

- 1942: R. Courant divided a domain into finite subregions

- 1956: J. Turner, R. W. Clough, H. C. Martin, and L. J. Topp's paper on the "Stiffness and Deflection of Complex Structures" introduces the name "finite-element method" and is widely recognized as the first comprehensive treatment of the method as it is known today

Basic Concepts

In this section, we are going to introduce some concepts from solid mechanics which will be useful for better understanding of this course. It is presumed that the readers have some basic knowledge of linear algebra and solid mechanics.

In solid mechanics, each phase of a material is considered to be continuum, that is, there is no discontinuity in the material. Thus, in this course individual fibres and the matrix of a lamina/composite are considered to be continuum. Further, this results in saying that heterogeneous composite is also a continuum.

Concept of Tensors

Tensors are physical entities whose components are the coefficients of a linear relationship between vectors.

The list of some of the tensors used in this course is given in Table below.

Table: List of some of the tensor quantities

	Quantity	Live subscripts
α	Scalar (zeroth order tensor)	0
v_i	Vector (first order tensor)	1
$\sigma_{ij}, \varepsilon_{ij}$	Second order tensor	2
C_{ijkl}	Fourth order tensor	4

It is often needed to transform a tensorial quantity from one coordinate system to another coordinate system. This transformation of a tensor is done using direction cosines of the angle measured from initial coordinate system to final coordinate system. Let us use axes x_i as the initial coordinate axes and x_i' as the final coordinate axes (denoted here by symbol prime). Now, we need to find the direction cosines (denoted here by a_{ij}) for this transformation relation. Let us use the convention for direction cosines that the first subscript (that is, i) of a_{ij} corresponds to the initial axes and the second subscript (that is, j) corresponds to final axes. The direction cosine correspondence with this convention in *3D* Cartesian coordinate system is given in Table. The corresponding Cartesian coordinate systems are shown in Figure.

Table: Direction cosines for 3D Cartesian coordinate system

From/To	x_1'	x_2'	x_3'
x_1	a_{11}	a_{12}	a_{13}
x_2	a_{21}	a_{22}	a_{23}
x_3	a_{31}	a_{32}	a_{33}

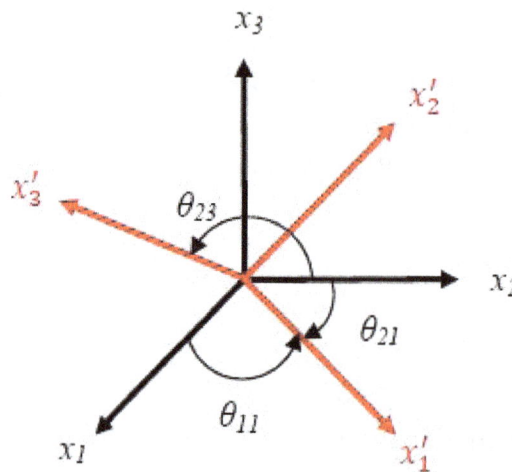

Rectangular or Cartesian coordinate systems

Let us derive the direction cosines for a transformation in a plane. Let the coordinate axes x_1-x_2 (that is, plane 1-2) are rotated about the third axis x_3 by an angle θ as shown in Figure. Thus, from the figure it is easy to see that $\theta_{11} = \theta_{22} = \theta$. A careful observation of the figure shows that the angle between x_1 and x_2' is not the same as the angle between x_2 and x_1'. It means that the direction cosines $a_{12} \neq a_{21}$.

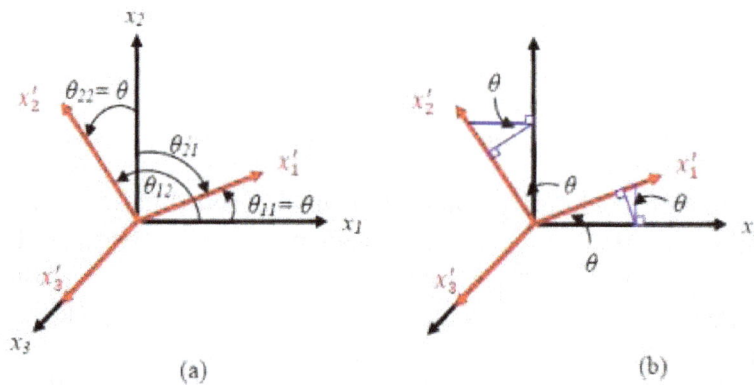

Transformation about x_3 axis

Now, we will find all the direction cosines. The list is given below.

$$a_{11} = \cos \theta_{11} = \cos \theta,$$

$$a_{12} = \cos \theta_{12} = \cos(90° + \theta) = -\sin \theta,$$

$$a_{13} = \cos 90° = 0,$$

$$a_{21} = \cos \theta_{21} = \sin \theta$$

$$a_{22} = \cos \theta_{22} = \cos \theta$$

$$a_{23} = \cos 90° = 0$$

$$a_{31} = \cos 90° = 0$$

$$a_{32} = \cos 90° = 0$$

$$a_{33} = \cos 0° = 1$$

The above can be written in a matrix form as

$$a_{ij} = \begin{bmatrix} \cos \theta & -\sin \theta & 0 \\ \sin \theta & \cos \theta & 0 \\ 0 & 0 & 1 \end{bmatrix} \quad (1)$$

The matrix of direction cosines given above in Eq. (1) is also written using short forms for $\cos \theta = m$ and $\sin \theta = n$. Then Equation (1) becomes

$$a_{ij} = \begin{bmatrix} m & -n & 0 \\ n & m & 0 \\ 0 & 0 & 1 \end{bmatrix} \quad (2)$$

Note: Some of the books and research articles also use $\cos \theta = c$ and $\sin \theta = s$.

Note: This matrix is also called Rotation Matrix.

Note: The above direction cosine matrix can be obtained from the relation between unrotated and rotated coordinates. For the transformation shown in Figure (a) one can write this relation using the geometrical relations shown in Figure (b) as

$$x_1' = x_1 \cos \theta_{11} + x_2 \sin \theta_{22}$$

$$x_2' = -x_1 \sin \theta_{11} + x_2 \cos \theta_{22}$$

$$x_3' = x_3$$

Now the direction cosines are given by the following relation:

$$a_{ji} = \frac{\partial x_i'}{\partial x_j}$$

Now we will use the direction cosines to transform a vector, a second order tensor and a fourth order tensor from initial coordinate (unprimed) system to a vector, a second order tensor and a fourth order tensor in final coordinate (primed) system.

First, let us do it for a vector. Let P_i and P_i' denote the components of a vector P in unprimed and primed coordinate axes. Then the components of this vector in rotated coordinate system are given in terms of components in unrotated coordinate system and corresponding direction cosines as

$$P_i' = a_{ji}P_j \quad (3)$$

Now, putting the direction cosines in terms of angles and summing over the repeated index j (=1, 2, 3) in Equation (3) we get

$$P'_i = a_{1i}P_1 + \alpha_{2i}P_2 + \alpha_{3i}P_3$$

$$P'_i = \cos\theta_{1i}P_1 + \cos\theta_{2i}P_2 + \cos\theta_{3i}P_3 \tag{4}$$

Let us assume that, the unprimed and primed coordinate systems are as shown in Figure (Transformation about x3 axis). The transformation matrix for this rotation is given in Equation (1). Then, the components P'_i can be given as

$$P'_1 = \cos\theta\, P_1 + \sin\theta\, P_2$$

$$P'_2 = -\sin\theta\, P_1 + \cos\theta\, P_2$$

$$P'_2 \quad P_3$$

Note: In two dimensional case, the above transformation is written as

$$P'_i = a_{ji}P_j = a_{1i}P_1 + \alpha_{2i}P_2 = \cos\theta_{1i}\, P_1 + \cos\theta_{2i}P_2$$

Equation (3) can also be written in an inverted form to give the components P_i in unrotated axes in terms of components P'_i in rotated axes system as

$$P_i = a_{ji}^{-1}P'_j \tag{5}$$

The rotation matrix a_{ij} in Equation (2) has a property that

$$a_{ji}^{-1} = a_{ij} = \left(a_{ji}\right)^T \tag{6}$$

Now, we will extend the concept to transform a second order tensor. Let us transform the stress tensor σ_{ij} as follows

$$\sigma'_{ij} = a_{ki}a_{lj}\sigma_{kl}$$

$$\sigma'_{ij} = a_{1i}a_{1j}a_{11} + a_{1i}a_{2j}\sigma_{12} + a_{1i}a_{3j}\sigma_{13}$$

$$+ a_{2i}a_{1j}\sigma_{21} + a_{2i}a_{2j}\sigma_{22} + a_{2i}a_{3j}\sigma_{23}$$

$$+ a_{3i}a_{1j}\sigma_{31} + a_{3i}a_{2j}\sigma_{32} + a_{3i}a_{3j}\sigma_{33}$$

$$\sigma'_{ij} = \cos\theta_{1i}\cos\theta_{1j}\sigma_{11} + \cos\theta_{1i}\cos\theta_{2j}\,\sigma_{12} + \cos\theta_{1i}\cos\theta_{3j}\sigma_{13}$$

$$+ \cos\theta_{2i}\cos\theta_{1j}\sigma_{21} + \cos\theta_{2i}\cos\theta_{2j}\,\sigma_{22} + \cos\theta_{2i}\cos\theta_{3j}\sigma_{23}$$

$$+ \cos\theta_{3i}\cos\theta_{1j}\sigma_{31} + \cos\theta_{3i}\cos\theta_{2j}\,\sigma_{32} + \cos\theta_{3i}\cos\theta_{3j}\sigma_{33} \tag{7}$$

The transformation of a fourth order tensor C_{ijkl} is given as

$$C'_{ijkl} = a_{pi}a_{qj}a_{rk}a_{sl}\, C_{pqrs} \tag{8}$$

The readers are suggested to write the final form of Equation (8) using similar procedure used to get the last of Equation (7).

Deformation of a Body

When a deformable body is subjected to external forces, a body may translate, rotate and deform as well. Thus, after deformation the body occupies a new region. The initial region occupied by the body is called Reference Configuration and the new region occupied by the body after translation, rotation and deformation is called Deformed Configuration. Let us consider a point P in reference configuration. Its position with respect to origin of a reference axes system (r) is shown in Figure. The point P occupies a new position P' and its position vector r' is also given.

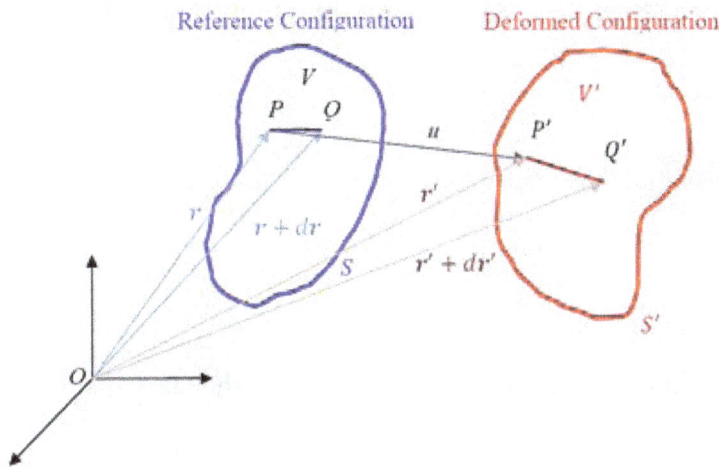

Reference and deformed configurations

The deformation map is defined as

$$r' = r'(r)$$ (9)

Thus, deformation map is a vector valued function. Similarly, for deformation of a point Q to Q', we can write

$$r' + dr' = r'(r + dr)$$ (10)

We can find the deformation dr' as

$$dr' = r'(r + dr) - r'(r) \approx (\nabla r')dr$$ (11)

where $(\nabla r') = F \dfrac{\partial r'}{\partial r}$ is called *Deformation Gradient.* In component form, one can write

$$dr'_i = F_{ij} dr_j$$ (12)

Now, let us give the deformation map for the displacement of a point. Let us consider the point P in reference configuration again. It undergoes a deformation $u(r)$ and occupies a new position P'. Thus, we can write this deformation as follows

$$u(r) = r' - r \text{ or } r' = r + u(r)$$

(13)

This gives us the deformation gradient as

$$F = \frac{\partial(r')}{\partial(r)} = I + \nabla u$$

(14)

or in component form

$$F_{ij} = \delta_{ij} + u_{ij}$$

(15)

Now, we will define strain tensor. We are going to find $|dr'|^2 - |dr|^2$. We know that $|dr'|^2 = dr' \cdot dr' = dr \cdot F^T F \, dr$. Thus,

$$
\begin{aligned}
|dr'|^2 - |dr'|^2 &= dr \cdot F^T F \, dr = dr \cdot dr \\
&= dr \cdot F^T F \, dr - dr \cdot I \, dr \\
&= dr \cdot (F^T F - I) dr \\
&= dr \cdot 2E \, dr
\end{aligned}
$$

(16)

where E is Lagrangian Strain Tensor. Now using the last two of Equation (16) for $E = \frac{1}{2}(F^T F - I)$ we get,

$$E = \frac{1}{2}(F^T F - I) = \frac{1}{2}\left[I + (\nabla u)^T (I + \nabla u) - I \right] = \frac{1}{2}\left[\nabla u + (\nabla u)^T + (\nabla u)^T \nabla u \right]$$

(17)

This equation can be written in index form as

$$\varepsilon_{ij} = \frac{1}{2}\left(u_{i,j} + u_{j,i} + u_{k,i} u_{k,j} \right)$$

(18)

where $u_{i,j}$ is given as $u_{i,j} = \frac{\partial u_i}{\partial x_j}$. Thus, the strain components are nonlinear in u_i. Here, $u_i = (u_1, u_2, u_3) = (u, v, w)$ are the displacement components in three directions. For example, let us write the expanded form of strain components ε_{11} and ε_{23}.

$$
\begin{aligned}
\varepsilon_{11} &= \frac{1}{2}\left[\frac{\partial u_1}{\partial x_1} + \frac{\partial u_1}{\partial x_1} + \left(\frac{\partial u_1}{\partial x_1}\right)^2 + \left(\frac{\partial u_2}{\partial x_1}\right)^2 + \left(\frac{\partial u_3}{\partial x_1}\right)^2 \right] \\
&= \frac{\partial u_1}{\partial x_1} + \frac{1}{2}\left[\left(\frac{\partial u_1}{\partial x_1}\right)^2 + \left(\frac{\partial u_2}{\partial x_1}\right)^2 + \left(\frac{\partial u_3}{\partial x_1}\right)^2 \right]
\end{aligned}
$$

(19)

Similarly,

$$\varepsilon_{23} = \frac{1}{2}\left[\frac{\partial u_2}{\partial x_3} + \frac{\partial u_3}{\partial x_2} + \frac{\partial u_1}{\partial x_2}\frac{\partial u_1}{\partial x_3} + \frac{\partial u_2}{\partial x_2}\frac{\partial u_2}{\partial x_3} + \frac{\partial u_3}{\partial x_2}\frac{\partial u_3}{\partial x_3}\right] \tag{20}$$

The readers should observe that from the definition of strain tensor in Equation (18), the strain tensor is symmetric (that is, $\varepsilon_{ij} = \varepsilon_{ji}$). If the gradients of the displacements are very small the product terms in Equation (18) can be neglected. Then, the resulting strain tensor (called Infinitesimal Strain Tensor) is given as

$$\varepsilon_{ij} = \frac{1}{2}\left(u_{i,j} + u_{j,i}\right) \tag{21}$$

The individual strain components are given as

$$\varepsilon_{11} = \frac{\partial u_1}{\partial x_1}, \quad \varepsilon_{12} = \varepsilon_{21} = \frac{1}{2}\left(\frac{\partial u_1}{\partial x_2} + \frac{\partial u_2}{\partial x_1}\right),$$

$$\varepsilon_{22} = \frac{\partial u_2}{\partial x_2}, \quad \varepsilon_{13} = \varepsilon_{31} = \frac{1}{2}\left(\frac{\partial u_1}{\partial x_3} + \frac{\partial u_3}{\partial x_1}\right),$$

$$\varepsilon_{33} = \frac{\partial u_3}{\partial x_3}, \quad \varepsilon_{23} = \varepsilon_{32} = \frac{1}{2}\left(\frac{\partial u_2}{\partial x_3} + \frac{\partial u_3}{\partial x_2}\right) \tag{22}$$

The readers are very well versed with these definitions. This strain tensor can be written in matrix form as

$$[\varepsilon] = \begin{bmatrix} \varepsilon_{11} & \varepsilon_{12} & \varepsilon_{13} \\ \varepsilon_{21} & \varepsilon_{22} & \varepsilon_{23} \\ \varepsilon_{31} & \varepsilon_{32} & \varepsilon_{33} \end{bmatrix} \tag{23}$$

Note: The shear strain components mentioned above are tensorial components. In actual practice, *engineering shear strains* (which are measured from laboratory tests) are used. These are denoted by γ_{ij}. The relation between tensorial and engineering shear strain components is

$$\gamma_{ij} = 2\varepsilon_{ij} \tag{24}$$

The engineering shear strain components are given as follows:

$$\gamma_{12} = \gamma_{21} = \frac{\partial u_1}{\partial x_2} + \frac{\partial u_2}{\partial x_1}$$

$$\gamma_{13} = \gamma_{31} = \frac{\partial u_1}{\partial x_3} + \frac{\partial u_3}{\partial x_1} \tag{25}$$

$$\gamma_{23} = \gamma_{32} = \frac{\partial u_2}{\partial x_3} + \frac{\partial u_3}{\partial x_2}$$

Using the engineering shear strain components, the strain tensor can be written in matrix form as

$$[\varepsilon] = \begin{bmatrix} \varepsilon_{11} & \dfrac{\gamma_{12}}{2} & \dfrac{\gamma_{13}}{2} \\ \dfrac{\gamma_{21}}{2} & \varepsilon_{22} & \dfrac{\gamma_{23}}{2} \\ \dfrac{\gamma_{31}}{2} & \dfrac{\gamma_{32}}{2} & \varepsilon_{33} \end{bmatrix} \tag{26}$$

Stress

Now, we will introduce the concept of stress. The components of stress at a point (also called State of Stress) are (in the limit) the forces per unit area which are acting on three mutually perpendicular planes passing through this point. This is represented in Figure. Stress tensor is a second order tensor and denoted as σ_{ij}. In this notation, the first subscript corresponds to the direction of the normal to the plane and the second subscript corresponds to the direction of the stress. For example, σ_{23} denotes the stress component acting on a plane which is perpendicular to direction 2 and stress is acting in direction 3. The tensile normal stress components $(\sigma_{11}, \sigma_{22}, \sigma_{33})$ are positive. The shear stress components $(i \neq j)$ are defined to be positive when the normal to the plane and the direction of the stress component are either both positive or both negative.

The readers should note that the state of stress shown in Figure represents all stress components in positive sense. In this figure, the stress components are shown on positive faces only.

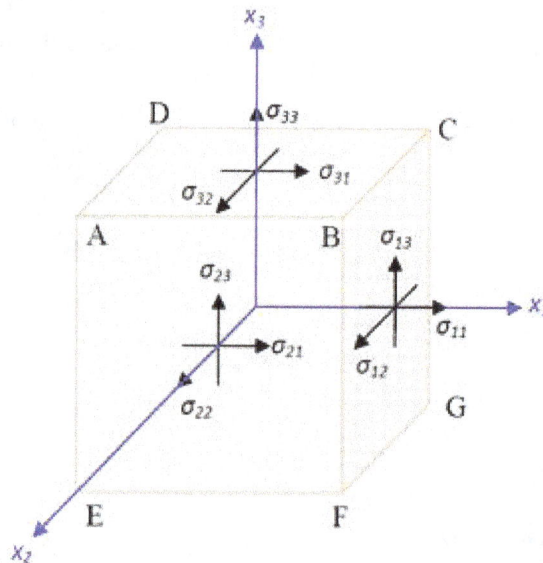

State of stress at a point

The stress tensor can be written in matrix form as follows:

$$[\sigma] = \begin{bmatrix} \sigma_{11} & \sigma_{12} & \sigma_{13} \\ \sigma_{21} & \sigma_{22} & \sigma_{23} \\ \sigma_{31} & \sigma_{32} & \sigma_{33} \end{bmatrix} \tag{27}$$

In general, instead of using global 1-2-3 coordinate system, x-y-z global coordinate system is used. Further, the shear stress components are shown using notation τ_{ij}. Thus, the stress tensor in this case can be written as

$$[\sigma] = \begin{bmatrix} \sigma_{xx} & \tau_{xy} & \tau_{xz} \\ \tau_{yx} & \sigma_{yy} & \tau_{yz} \\ \tau_{zx} & \tau_{zy} & \sigma_{zz} \end{bmatrix} \tag{28}$$

Note: The stress tensor will be symmetric, that is $\sigma_{ij} = \sigma_{ji}$ only when there are no distributed moments in the body. The readers are suggested to read more on this from any standard solid mechanics book. In this entire course, we will deal with symmetric stress-tensor.

Equilibrium Equations

The equilibrium equations for a body to be in static equilibrium at a point are given in index notations as

$$\sigma_{ij,j} + b_i = 0 \tag{29}$$

where, b_i are the body forces per unit volume. If the body forces are absent, then the equilibrium equation becomes

$$\sigma_{ij,j} = 0 \tag{30}$$

The equilibrium equations, without body forces are written using xyz coordinates as follows:

$$\frac{\partial \sigma_{xx}}{\partial x} + \frac{\partial \sigma_{xy}}{\partial y} + \frac{\partial \sigma_{xz}}{\partial z} = 0$$

$$\frac{\partial \sigma_{yx}}{\partial x} + \frac{\partial \sigma_{yy}}{\partial y} + \frac{\partial \sigma_{yz}}{\partial z} = 0$$

$$\frac{\partial \sigma_{zx}}{\partial x} + \frac{\partial \sigma_{zy}}{\partial y} + \frac{\partial \sigma_{zz}}{\partial z} = 0 \tag{31}$$

Boundary Conditions

The boundary conditions are very essential to solve any problem in solid mechanics. The boundary conditions are specified on the surface of the body in terms of components of displacement or traction. However, the combination of displacement and traction components is also specified.

Figure shows a body, where the displacement as well as traction components are used to specify the boundary conditions.

We define traction vector T_i for any arbitrary point (for example, point P in Figure) on surface as a vector consisting of three stress components acting on the surface at same point. Here, the three stress components are normal stress σ_{nn} and shear stress σ_{nt} and σ_{ns}. The traction vector at this point is written as

$$T_i - \sigma_{ji} n_j \tag{32}$$

where n_i is the ith component of the unit normal to the surface at point P. For example, if this surface is perpendicular to axis 2, then $n_i = (0,1,0)$ and the components of traction acting at a point on this surface are given as follows

$$T_1 = \sigma_{21}$$
$$T_2 = \sigma_{22}$$
$$T_3 = \sigma_{23} \tag{33}$$

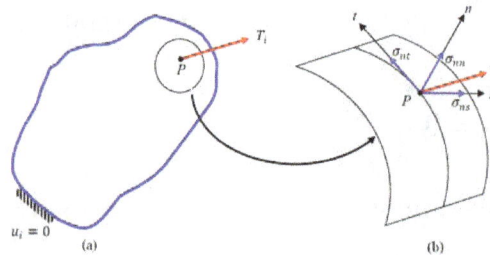

(a) A body showing displacement and traction boundary conditions,
(b) Traction vector at any arbitrary point P on the surface of a body

Constitutive Equations

The relationship between stress and strain is known as constitutive equation. The general form of this equation is

$$\sigma_{ij} = C_{ijkl}\,\varepsilon_{kl} \tag{34}$$

Here, C_{ijkl} are called elastic constants. This is also referred to as elastic moduli or elastic stiffnesses. This form of constitutive equation is known as generalized Hooke's law. Very soon, we will see this equation in detail for various material types.

The inverse of this equation can be written as

$$\varepsilon_{ij} = C_{ijkl}^{-1}\sigma_{kl} = S_{ijkl}\sigma_{kl} \tag{35}$$

where S_{ijkl} is known as compliance.

Plane Stress Problem

Plane stress problem corresponds to a situation where out of plane stress components are negligibly small. Thus, we can say that the state of stress is planar. The planar state of stress in x-y plane is shown in Figure. For the case shown in this figure, the normal and shear stress components in z directions, that is σ_{zz}, σ_{xz} and σ_{yz} are zero. Please note that the state of stress shown in this figure assumes the stress symmetry.

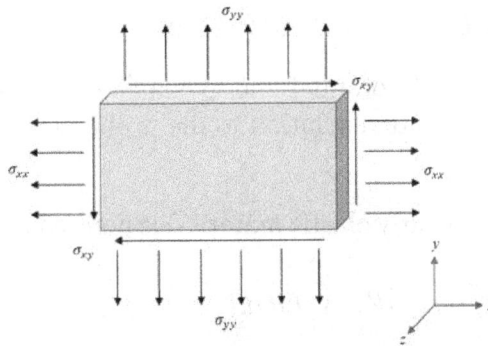

Plane stress problem

Note: A careful observation for strain components in z direction ($\in_{zz}, \in_{xz}$ and $\in_{yz}$) reveals that these need not be zero. This is a common mistake made by many readers. The magnitude of these strain components can be found with the help of constitutive equation given in Equation (34).

For plane stress problem the equilibrium equations take the following form

$$\frac{\partial \sigma_{xx}}{\partial x} + \frac{\partial \sigma_{xy}}{\partial y} = 0$$

$$\frac{\partial \sigma_{xx}}{\partial x} + \frac{\partial \sigma_{xy}}{\partial y} = 0 \tag{36}$$

Plane Strain Problem

Plane strain problem corresponds to a condition where all the out of plane strain components are negligibly small. Here, we denote $\in_{xz}, \in_{yz}$ and $\in_{zz}$ as out of plane strain components. The readers are again cautioned to note that the out of plane stress components need not be zero. These depend upon the constitutive equation. Further, the equilibrium equation is same as Equation (36) and $\sigma_{zz} = f(x, y)$.

Principles from Work and Energy

Strain Energy Density

The strain energy stored in a body per unit volume is called as strain energy density. In the absence of internal energy, the strain energy density for a linearly elastic body is given as

$$W = \frac{1}{2}\sigma_{ij}\in_{ij} \tag{37}$$

The expanded form of the above equation using symmetry of stress and strain components is

$$W = \frac{1}{2}\left(\sigma_{11}\varepsilon_{11} + \sigma_{11}\varepsilon_{22} + \sigma_{33}\varepsilon_{33} + 2\sigma_{13}\varepsilon_{13} + 2\sigma_{23}\varepsilon_{23}\right)$$

(38)

The readers should note that strain energy density is a scalar quantity. Further, it is a positive definite quantity.

Principle of Minimum of Total Potential Energy

The principle of minimum of total potential energy states that of all possible kinematically admissible displacement fields, the actual solution to the problem is one which minimizes the total potential energy (Π).

The total potential energy (for linearly elastic material) is defined as

$$\Pi = \frac{1}{2}\int_V C_{ijkl}\varepsilon_{ij}\varepsilon_{kl}\,dV - \int_S T_i u_i\,dS$$

(39)

Note: The kinematically admissible displacement field is a single valued and continuous displacement field that satisfies the displacement boundary condition.

Principle of Minimum of Total Complementary Potential Energy

The principle of minimum of total complementary potential energy states that of all possible statically admissible stress fields, the actual solution to the problem is one which minimizes the total complementary potential energy (Π^*).

The total complementary potential energy (for linearly elastic material) is defined as

$$\Pi^* = \frac{1}{2}\int_V S_{ijkl}\sigma_{ij}\sigma_{kl}\,dV - \int_S T_i^{\circ} u_i\,dS$$

(40)

Note: The statically admissible stress field is one that satisfies both equilibrium equations and traction boundary condition.

Strength of Materials

Strength of materials, also called mechanics of materials, is a subject which deals with the behavior of solid objects subject to stresses and strains. The complete theory began with the consideration of the behavior of one and two dimensional members of structures, whose states of stress can be approximated as two dimensional, and was then generalized to three dimensions to develop a more complete theory of the elastic and plastic behavior of materials. An important founding pioneer in mechanics of materials was Stephen Timoshenko.

The study of strength of materials often refers to various methods of calculating the stresses and strains in structural members, such as beams, columns, and shafts. The methods employed to predict the response of a structure under loading and its susceptibility to various failure modes takes into account the properties of the materials such as its yield strength, ultimate strength, Young's modulus, and Poisson's ratio; in addition the mechanical element's macroscopic properties (geometric properties), such as its length, width, thickness, boundary constraints and abrupt changes in geometry such as holes are considered.

Definition

In materials science, the strength of a material is its ability to withstand an applied load without failure or plastic deformation. The field of strength of materials deals with forces and deformations that result from their acting on a material. A load applied to a mechanical member will induce internal forces within the member called stresses when those forces are expressed on a unit basis. The stresses acting on the material cause deformation of the material in various manners. Deformation of the material is called strain when those deformations too are placed on a unit basis. The applied loads may be axial (tensile or compressive), or rotational (strength shear). The stresses and strains that develop within a mechanical member must be calculated in order to assess the load capacity of that member. This requires a complete description of the geometry of the member, its constraints, the loads applied to the member and the properties of the material of which the member is composed. With a complete description of the loading and the geometry of the member, the state of stress and of state of strain at any point within the member can be calculated. Once the state of stress and strain within the member is known, the strength (load carrying capacity) of that member, its deformations (stiffness qualities), and its stability (ability to maintain its original configuration) can be calculated. The calculated stresses may then be compared to some measure of the strength of the member such as its material yield or ultimate strength. The calculated deflection of the member may be compared to a deflection criteria that is based on the member's use. The calculated buckling load of the member may be compared to the applied load. The calculated stiffness and mass distribution of the member may be used to calculate the member's dynamic response and then compared to the acoustic environment in which it will be used.

Material strength refers to the point on the engineering stress–strain curve (yield stress) beyond which the material experiences deformations that will not be completely reversed upon removal of the loading and as a result the member will have a permanent deflection. The ultimate strength refers to the point on the engineering stress–strain curve corresponding to the stress that produces fracture.

Types of Loadings

- Transverse loading — Forces applied perpendicular to the longitudinal axis of a member. Transverse loading causes the member to bend and deflect from its original position, with internal tensile and compressive strains accompanying the change in curvature of the member. Transverse loading also induces shear forces that cause shear deformation of the material and increase the transverse deflection of the member.

- Axial loading — The applied forces are collinear with the longitudinal axis of the member. The forces cause the member to either stretch or shorten.

- Torsional loading — Twisting action caused by a pair of externally applied equal and oppositely directed force couples acting on parallel planes or by a single external couple applied to a member that has one end fixed against rotation.

Stress Terms

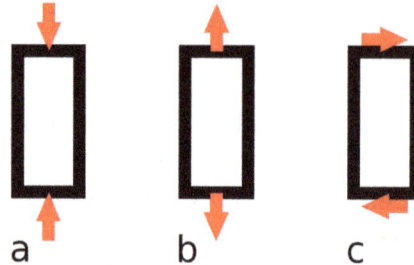

A material being loaded in a) compression, b) tension, c) shear.

Uniaxial stress is expressed by

$$\sigma = \frac{F}{A}$$

where F is the force [N] acting on an area A [m^2]. The area can be the undeformed area or the deformed area, depending on whether engineering stress or true stress is of interest.

- *Compressive stress* (or compression) is the stress state caused by an applied load that acts to reduce the length of the material (compression member) along the axis of the applied load, it is in other words a stress state that causes a squeezing of the material. A simple case of compression is the uniaxial compression induced by the action of opposite, pushing forces. Compressive strength for materials is generally higher than their tensile strength. However, structures loaded in compression are subject to additional failure modes, such as buckling, that are dependent on the member's geometry.

- *Tensile stress* is the stress state caused by an applied load that tends to elongate the material along the axis of the applied load, in other words the stress caused by *pulling* the material. The strength of structures of equal cross sectional area loaded in tension is independent of shape of the cross section. Materials loaded in tension are susceptible to stress concentrations such as material defects or abrupt changes in geometry. However, materials exhibiting ductile behavior (most metals for example) can tolerate some defects while brittle materials (such as ceramics) can fail well below their ultimate material strength.

- *Shear stress* is the stress state caused by the combined energy of a pair of opposing forces acting along parallel lines of action through the material, in other words the stress caused by faces of the material *sliding* relative to one another. An example is cutting paper with scissors or stresses due to torsional loading.

Strength Terms

Mechanical properties of materials include the yield strength, tensile strength, fatigue strength, crack resistance, and other characteristics.

- *Yield strength* is the lowest stress that produces a permanent deformation in a material. In some materials, like aluminium alloys, the point of yielding is difficult to identify, thus it is usually defined as the stress required to cause 0.2% plastic strain. This is called a 0.2% proof stress.

- *Compressive strength* is a limit state of compressive stress that leads to failure in a material in the manner of ductile failure (infinite theoretical yield) or brittle failure (rupture as the result of crack propagation, or sliding along a weak plane).

- *Tensile strength* or *ultimate tensile strength* is a limit state of tensile stress that leads to tensile failure in the manner of ductile failure (yield as the first stage of that failure, some hardening in the second stage and breakage after a possible "neck" formation) or brittle failure (sudden breaking in two or more pieces at a low stress state). Tensile strength can be quoted as either true stress or engineering stress, but engineering stress is the most commonly used.

- *Fatigue strength* is a measure of the strength of a material or a component under cyclic loading, and is usually more difficult to assess than the static strength measures. Fatigue strength is quoted as stress amplitude or stress range ($\Delta\sigma = \sigma_{max} - \sigma_{min}$), usually at zero mean stress, along with the number of cycles to failure under that condition of stress.

- *Impact strength,* is the capability of the material to withstand a suddenly applied load and is expressed in terms of energy. Often measured with the Izod impact strength test or Charpy impact test, both of which measure the impact energy required to fracture a sample. Volume, modulus of elasticity, distribution of forces, and yield strength affect the impact strength of a material. In order for a material or object to have a high impact strength the stresses must be distributed evenly throughout the object. It also must have a large volume with a low modulus of elasticity and a high material yield strength.

Strain (Deformation) Terms

- *Deformation* of the material is the change in geometry created when stress is applied (as a result of applied forces, gravitational fields, accelerations, thermal expansion, etc.). Deformation is expressed by the displacement field of the material.

- *Strain* or *reduced deformation* is a mathematical term that expresses the trend of the deformation change among the material field. Strain is the deformation per unit length. In the case of uniaxial loading the displacements of a specimen (for example a bar element) lead to a calculation of strain expressed as the quotient of the displacement and the original length of the specimen. For 3D displacement fields it is expressed as derivatives of displacement functions in terms of a second order tensor (with 6 independent elements).

- *Deflection* is a term to describe the magnitude to which a structural element is displaced when subject to an applied load.

Stress–strain Relations

- *Elasticity* is the ability of a material to return to its previous shape after stress is released. In many materials, the relation between applied stress is directly proportional to the resulting strain (up to a certain limit), and a graph representing those two quantities is a straight line.

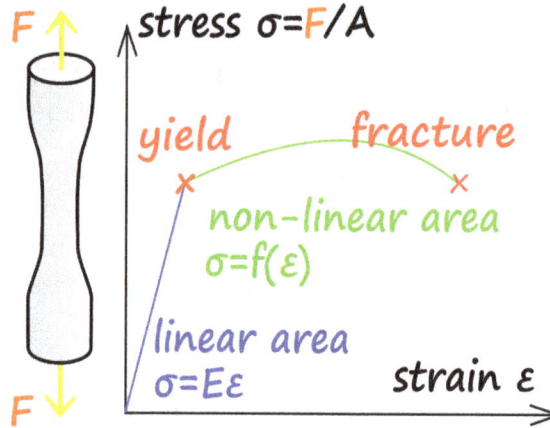

Basic static response of a specimen under tension

The slope of this line is known as Young's modulus, or the "modulus of elasticity." The modulus of elasticity can be used to determine the stress–strain relationship in the linear-elastic portion of the stress–strain curve. The linear-elastic region is either below the yield point, or if a yield point is not easily identified on the stress–strain plot it is defined to be between 0 and 0.2% strain, and is defined as the region of strain in which no yielding (permanent deformation) occurs.

- *Plasticity* or plastic deformation is the opposite of elastic deformation and is defined as unrecoverable strain. Plastic deformation is retained after the release of the applied stress. Most materials in the linear-elastic category are usually capable of plastic deformation. Brittle materials, like ceramics, do not experience any plastic deformation and will fracture under relatively low strain, while ductile materials such as metallics, lead, or polymers will plasticly deform much more before a fracture initiation.

Consider the difference between a carrot and chewed bubble gum. The carrot will stretch very little before breaking. The chewed bubble gum, on the other hand, will plastically deform enormously before finally breaking.

Design Terms

Ultimate strength is an attribute related to a material, rather than just a specific specimen made of the material, and as such it is quoted as the force per unit of cross section area (N/m^2). The ultimate strength is the maximum stress that a material can withstand before it breaks or weakens. For example, the ultimate tensile strength (UTS) of AISI 1018 Steel is 440 MN/m^2. In general, the SI unit of stress is the pascal, where 1 Pa = 1 N/m^2. In Imperial units, the unit of stress is given as lbf/in^2 or pounds-force per square inch. This unit is often abbreviated as psi. One thousand psi is abbreviated ksi.

A Factor of safety is a design criteria that an engineered component or structure must achieve.

$FS = UTS / R$, where FS: the factor of safety, R: The applied stress, and UTS: ultimate stress (psi or N/m²)

Margin of Safety is also sometimes used to as design criteria. It is defined MS = Failure Load/(Factor of Safety × Predicted Load) – 1.

For example, to achieve a factor of safety of 4, the allowable stress in an AISI 1018 steel component can be calculated to be $R = UTS / FS$ = 440/4 = 110 MPa, or R = 110×10⁶ N/m². Such allowable stresses are also known as "design stresses" or "working stresses."

Design stresses that have been determined from the ultimate or yield point values of the materials give safe and reliable results only for the case of static loading. Many machine parts fail when subjected to a non steady and continuously varying loads even though the developed stresses are below the yield point. Such failures are called fatigue failure. The failure is by a fracture that appears to be brittle with little or no visible evidence of yielding. However, when the stress is kept below "fatigue stress" or "endurance limit stress", the part will endure indefinitely. A purely reversing or cyclic stress is one that alternates between equal positive and negative peak stresses during each cycle of operation. In a purely cyclic stress, the average stress is zero. When a part is subjected to a cyclic stress, also known as stress range (Sr), it has been observed that the failure of the part occurs after a number of stress reversals (N) even if the magnitude of the stress range is below the material's yield strength. Generally, higher the range stress, the fewer the number of reversals needed for failure.

Failure Theories

There are four failure theories: maximum shear stress theory, maximum normal stress theory, maximum strain energy theory, and maximum distortion energy theory. Out of these four theories of failure, the maximum normal stress theory is only applicable for brittle materials, and the remaining three theories are applicable for ductile materials. Of the latter three, the distortion energy theory provides most accurate results in majority of the stress conditions. The strain energy theory needs the value of Poisson's ratio of the part material, which is often not readily available. The maximum shear stress theory is conservative. For simple unidirectional normal stresses all theories are equivalent, which means all theories will give the same result.

- Maximum Shear stress Theory — This theory postulates that failure will occur if the magnitude of the maximum shear stress in the part exceeds the shear strength of the material determined from uniaxial testing.

- Maximum normal stress theory — This theory postulates that failure will occur if the maximum normal stress in the part exceeds the ultimate tensile stress of the material as determined from uniaxial testing. This theory deals with brittle materials only. The maximum tensile stress should be less than or equal to ultimate tensile stress divided by factor of safety. The magnitude of the maximum compressive stress should be less than ultimate compressive stress divided by factor of safety.

- Maximum strain energy theory — This theory postulates that failure will occur when the strain energy per unit volume due to the applied stresses in a part equals the strain energy per unit volume at the yield point in uniaxial testing.

- Maximum distortion energy theory — This theory is also known as shear energy theory or von Mises-Hencky theory. This theory postulates that failure will occur when the distortion energy per unit volume due to the applied stresses in a part equals the distortion energy per unit volume at the yield point in uniaxial testing. The total elastic energy due to strain can be divided into two parts: one part causes change in volume, and the other part causes change in shape. Distortion energy is the amount of energy that is needed to change the shape.

- Fracture mechanics was established by Alan Arnold Griffith and George Rankine Irwin. This important theory is also known as numeric conversion of toughness of material in the case of crack existence.

- Fractology was proposed by Takeo Yokobori because each fracture laws including creep rupture criterion must be combined nonlinearly.

A material's strength is dependent on its microstructure. The engineering processes to which a material is subjected can alter this microstructure. The variety of strengthening mechanisms that alter the strength of a material includes work hardening, solid solution strengthening, precipitation hardening and grain boundary strengthening and can be quantitatively and qualitatively explained. Strengthening mechanisms are accompanied by the caveat that some other mechanical properties of the material may degenerate in an attempt to make the material stronger. For example, in grain boundary strengthening, although yield strength is maximized with decreasing grain size, ultimately, very small grain sizes make the material brittle. In general, the yield strength of a material is an adequate indicator of the material's mechanical strength. Considered in tandem with the fact that the yield strength is the parameter that predicts plastic deformation in the material, one can make informed decisions on how to increase the strength of a material depending its microstructural properties and the desired end effect. Strength is expressed in terms of the limiting values of the compressive stress, tensile stress, and shear stresses that would cause failure. The effects of dynamic loading are probably the most important practical consideration of the strength of materials, especially the problem of fatigue. Repeated loading often initiates brittle cracks, which grow until failure occurs. The cracks always start at stress concentrations, especially changes in cross-section of the product, near holes and corners at nominal stress levels far lower than those quoted for the strength of the material.

Plane Stress

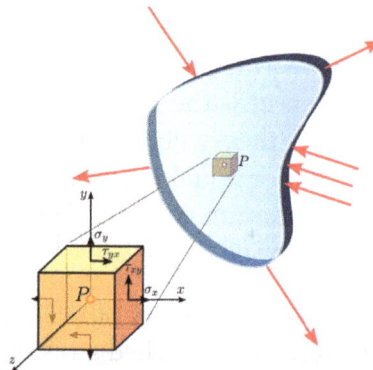

Plane stress state in a continuum.

In continuum mechanics, a material is said to be under plane stress if the stress vector is zero across a particular surface. When that situation occurs over an entire element of a structure, as is often the case for thin plates, the stress analysis is considerably simplified, as the stress state can be represented by a tensor of dimension 2 (representable as a 2 × 2 matrix rather than 3 × 3). A related notion, plane strain, is often applicable to very thick members.

Plane stress typically occurs in thin flat plates that are acted upon only by load forces that are parallel to them. In certain situations, a gently curved thin plate may also be assumed to have plane stress for the purpose of stress analysis. This is the case, for example, of a thin-walled cylinder filled with a fluid under pressure. In such cases, stress components perpendicular to the plate are negligible compared to those parallel to it.

In other situations, however, the bending stress of a thin plate cannot be neglected. One can still simplify the analysis by using a two-dimensional domain, but the plane stress tensor at each point must be complemented with bending terms.

Mathematical Definition

Mathematically, the stress at some point in the material is a plane stress if one of the three principal stresses (the eigenvalues of the Cauchy stress tensor) is zero. That is, there is Cartesian coordinate system in which the stress tensor has the form

$$\sigma = \begin{bmatrix} \sigma_{11} & 0 & 0 \\ 0 & \sigma_{22} & 0 \\ 0 & 0 & 0 \end{bmatrix} \equiv \begin{bmatrix} \sigma_x & 0 & 0 \\ 0 & \sigma_y & 0 \\ 0 & 0 & 0 \end{bmatrix}$$

For example, consider a rectangular block of material measuring 10, 40 and 5 cm along the x, y, and z, that is being stretched in the x direction and compressed in the y direction, by pairs of opposite forces with magnitudes 10 N and 20 N, respectively, uniformly distributed over the corresponding faces. The stress tensor inside the block will be

$$\sigma = \begin{bmatrix} 500\text{Pa} & 0 & 0 \\ 0 & -4000\text{Pa} & 0 \\ 0 & 0 & 0 \end{bmatrix}$$

More generally, if one chooses the first two coordinate axes arbitrarily but perpendicular to the direction of zero stress, the stress tensor will have the form

$$\sigma = \begin{bmatrix} \sigma_{11} & \sigma_{12} & 0 \\ \sigma_{21} & \sigma_{22} & 0 \\ 0 & 0 & 0 \end{bmatrix} \equiv \begin{bmatrix} \sigma_x & \tau_{xy} & 0 \\ \tau_{yx} & \sigma_y & 0 \\ 0 & 0 & 0 \end{bmatrix}$$

and can therefore be represented by a 2 × 2 matrix,

$$\sigma_{ij} = \begin{bmatrix} \sigma_{11} & \sigma_{12} \\ \sigma_{21} & \sigma_{22} \end{bmatrix} \equiv \begin{bmatrix} \sigma_x & \tau_{xy} \\ \tau_{yx} & \sigma_y \end{bmatrix}$$

Constitutive Equations

Plane Stress in Curved Surfaces

In certain cases, the plane stress model can be used in the analysis of gently curved surfaces. For example, consider a thin-walled cylinder subjected to an axial compressive load uniformly distributed along its rim, and filled with a pressurized fluid. The internal pressure will generate a reactive hoop stress on the wall, a normal tensile stress directed perpendicular to the cylinder axis and tangential to its surface. The cylinder can be conceptually unrolled and analyzed as a flat thin rectangular plate subjected to tensile load in one direction and compressive load in another other direction, both parallel to the plate.

Plane Strain (Strain Matrix)

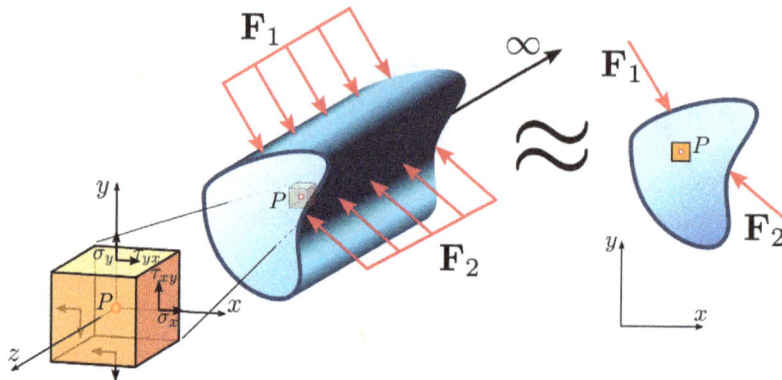

Plane strain state in a continuum.

If one dimension is very large compared to the others, the principal strain in the direction of the longest dimension is constrained and can be assumed as zero, yielding a plane strain condition. In this case, though all principal stresses are non-zero, the principal stress in the direction of the longest dimension can be disregarded for calculations. Thus, allowing a two dimensional analysis of stresses, e.g. a dam analyzed at a cross section loaded by the reservoir.

The corresponding strain tensor is:

$$\varepsilon_{ij} = \begin{bmatrix} \varepsilon_{11} & \varepsilon_{12} & 0 \\ \varepsilon_{21} & \varepsilon_{22} & 0 \\ 0 & 0 & \varepsilon_{33} \end{bmatrix}$$

in which the non-zero ε_{33} term arises from the Poisson's effect. This strain term can be temporarily removed from the stress analysis to leave only the in-plane terms, effectively reducing the analysis to two dimensions.

Stress Transformation in Plane Stress and Plane Strain

Consider a point P in a continuum under a state of plane stress, or plane strain, with stress components $(\sigma_x, \sigma_y, \tau_{xy})$ and all other stress components equal to zero. From static equilibrium of an infinitesimal material element at P, the normal stress σ_n and the shear stress τ_n on any plane perpendicular to the x - y plane passing through P with a unit vector $\mathbf{n}$ making an angle of θ with the horizontal, i.e. $\cos\theta$ is the direction cosine in the x direction, is given by:

$$\sigma_n = \frac{1}{2}(\sigma_x + \sigma_y) + \frac{1}{2}(\sigma_x - \sigma_y)\cos 2\theta + \tau_{xy}\sin 2\theta$$

$$\tau_n = -\frac{1}{2}(\sigma_x - \sigma_y)\sin 2\theta + \tau_{xy}\cos 2\theta$$

These equations indicate that in a plane stress or plane strain condition, one can determine the stress components at a point on all directions, i.e. as a function of θ, if one knows the stress components $(\sigma_x, \sigma_y, \tau_{xy})$ on any two perpendicular directions at that point. It is important to remember that we are considering a unit area of the infinitesimal element in the direction parallel to the θ - z plane.

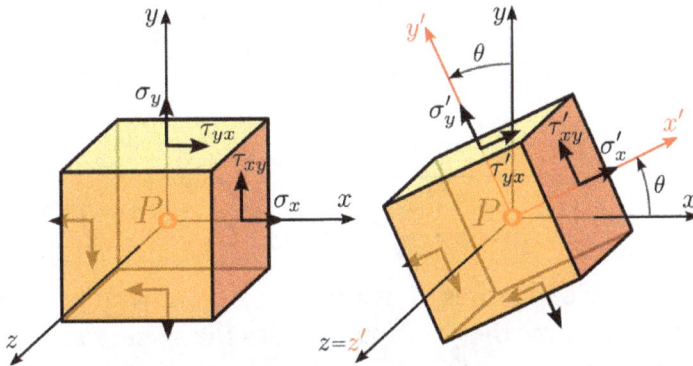

Stress transformation at a point in a continuum under plane stress conditions.

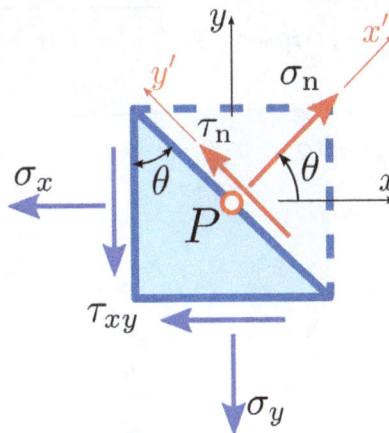

Stress components at a plane passing through a point in a continuum under plane stress conditions.

The principal directions, i.e., orientation of the planes where the shear stress components are zero, can be obtained by making the previous equation for the shear stress τ_n equal to zero. Thus we have:

$$\tau_n = -\frac{1}{2}(\sigma_x - \sigma_y)\sin 2\theta + \tau_{xy}\cos 2\theta = 0$$

and we obtain

$$\tan 2\theta_p = \frac{2\tau_{xy}}{\sigma_x - \sigma_y}$$

This equation defines two values θ_p which are $90°$ apart. The same result can be obtained by finding the angle θ which makes the normal stress σ_n a maximum, i.e. $\dfrac{d\sigma_n}{d\theta} = 0$ The principal stresses σ_1 and σ_2, or minimum and maximum normal stresses σ_{max} and σ_{min}, respectively, can then be obtained by replacing both values of θ_p into the previous equation for σ_n . This can be achieved by rearranging the equations for σ_n and τ_n, first transposing the first term in the first equation and squaring both sides of each of the equations then adding them. Thus we have

$$\left[\sigma_n - \tfrac{1}{2}(\sigma_x + \sigma_y)\right]^2 + \tau_n^2 = \left[\tfrac{1}{2}(\sigma_x - \sigma_y)\right]^2 + \tau_{xy}^2$$

$$(\sigma_n - \sigma_{avg})^2 + \tau_n^2 = R^2$$

where

$$R = \sqrt{\left[\tfrac{1}{2}(\sigma_x - \sigma_y)\right]^2 + \tau_{xy}^2} \quad \text{and} \quad \sigma_{avg} = \tfrac{1}{2}(\sigma_x + \sigma_y)$$

which is the equation of a circle of radius R centered at a point with coordinates $[\sigma_{avg}, 0]$, called Mohr's circle. But knowing that for the principal stresses the shear stress $\tau_n = 0$, then we obtain from this equation:

$$\sigma_1 = \sigma_{max} = \tfrac{1}{2}(\sigma_x + \sigma_y) + \sqrt{\left[\tfrac{1}{2}(\sigma_x - \sigma_y)\right]^2 + \tau_{xy}^2}$$

$$\sigma_2 = \sigma_{min} = \tfrac{1}{2}(\sigma_x + \sigma_y) - \sqrt{\left[\tfrac{1}{2}(\sigma_x - \sigma_y)\right]^2 + \tau_{xy}^2}$$

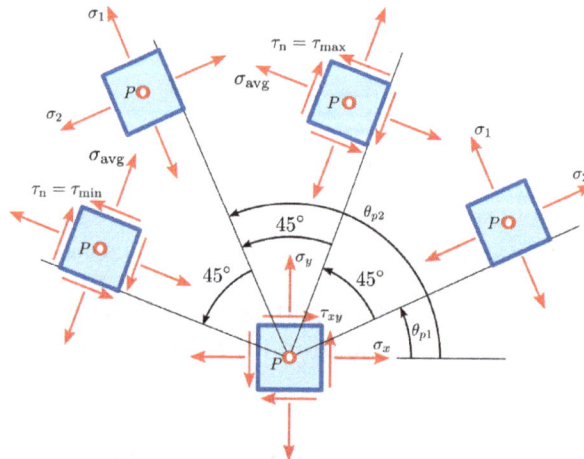

Transformation of stresses in two dimensions, showing the planes of action of principal stresses, and maximum and minimum shear stresses.

When $\tau_{xy} = 0$ the infinitesimal element is oriented in the direction of the principal planes, thus the stresses acting on the rectangular element are principal stresses: $\sigma_x = \sigma_1$ and $\sigma_y = \sigma_2$. Then the normal stress σ_n and shear stress τ_n as a function of the principal stresses can be determined by making $\tau_{xy} = 0$. Thus we have

$$\sigma_n = \frac{1}{2}(\sigma_1 + \sigma_2) + \frac{1}{2}(\sigma_1 - \sigma_2)\cos 2\theta$$

$$\tau_n = -\frac{1}{2}(\sigma_1 - \sigma_2)\sin 2\theta$$

Then the maximum shear stress τ_{max} occurs when $\sin 2\theta = 1$, i.e. $\theta = 45°$

$$\tau_{max} = \frac{1}{2}(\sigma_1 - \sigma_2)$$

Then the minimum shear stress τ_{min} occurs when $\sin 2\theta = -1$, i.e. $\theta = 135°$

$$\tau_{min} = -\frac{1}{2}(\sigma_1 - \sigma_2)$$

References

- Clifford Truesdell & Walter Noll; Stuart S. Antman, editor (2004). The Non-linear Field Theories of Mechanics. Springer. p. 4. ISBN 3-540-02779-3

- Jørgen Rammer (2007). Quantum Field Theory of Nonequilibrium States. Cambridge University Press. ISBN 978-0-521-87499-1

- Vitaliy Lomakin; Steinberg BZ; Heyman E; Felsen LB (2003). "Multiresolution Homogenization of Field and Network Formulations for Multiscale Laminate Dielectric Slabs" (PDF)

- Kay, J.M. (1985). Fluid Mechanics and Transfer Processes. Cambridge University Press. pp. 10 & 122–124. ISBN 9780521316248

- O. C. Zienkiewicz; Robert Leroy Taylor; J. Z. Zhu; Perumal Nithiarasu (2005). The Finite Element Method (Sixth ed.). Oxford UK: Butterworth-Heinemann. p. 550 ff. ISBN 0-7506-6321-9

Constitutive Relations: An Overview

Constitutive equations describe the relation between two physical quantities. It is used to derive the response of a material to external stimuli. Constitutive equations can either be phenomenological or can be derived from first principles. The topics discussed in the section are of great importance to broaden the existing knowledge on constitutive equations.

Constitutive Equation

In physics and engineering, a constitutive equation or constitutive relation is a relation between two physical quantities (especially kinetic quantities as related to kinematic quantities) that is specific to a material or substance, and approximates the response of that material to external stimuli, usually as applied fields or forces. They are combined with other equations governing physical laws to solve physical problems; for example in fluid mechanics the flow of a fluid in a pipe, in solid state physics the response of a crystal to an electric field, or in structural analysis, the connection between applied stresses or forces to strains or deformations.

Some constitutive equations are simply phenomenological; others are derived from first principles. A common approximate constitutive equation frequently is expressed as a simple proportionality using a parameter taken to be a property of the material, such as electrical conductivity or a spring constant. However, it is often necessary to account for the directional dependence of the material, and the scalar parameter is generalized to a tensor. Constitutive relations are also modified to account for the rate of response of materials and their non-linear behavior.

Mechanical Properties of Matter

The first constitutive equation (constitutive law) was developed by Robert Hooke and is known as Hooke's law. It deals with the case of linear elastic materials. Following this discovery, this type of equation, often called a "stress-strain relation" in this example, but also called a "constitutive assumption" or an "equation of state" was commonly used. Walter Noll advanced the use of constitutive equations, clarifying their classification and the role of invariance requirements, constraints, and definitions of terms like "material", "isotropic", "aeolotropic", etc. The class of "constitutive relations" of the form *stress rate = f (velocity gradient, stress, density)* was the subject of Walter Noll's dissertation in 1954 under Clifford Truesdell.

In modern condensed matter physics, the constitutive equation plays a major role.

Definitions

Quantity (common name/s)	(Common) symbol/s	Defining equation	SI units	Dimension
General stress, Pressure	P, σ	$\sigma = F / A$ F may be any perpendicular force applied to area A	Pa = N m^{-2}	[M] [T]$^{-2}$[L]$^{-1}$
General strain	ε	$\varepsilon = \Delta D / D$ D = dimension (length, area, volume) ΔK = change in material	dimensionless	dimensionless
General elastic modulus	E_{mod}	$E_{mod} = \sigma / \varepsilon$	Pa = N m^{-2}	[M] [T]$^{-2}$ [L]$^{-1}$
Young's modulus	E, Y	$Y = \;/(\Delta L / L)$	Pa = N m^{-2}	[M] [T] $^{-2}$[L]$^{-1}$
Shear modulus	G	$G = \Delta x / L$	Pa = N m^{-2}	[M] [T]$^{-2}$[L]$^{-1}$
Bulk modulus	K, B	$B = P / (\Delta V / V)$	Pa = N m^{-2}	[M] [T]$^{-2}$[L]$^{-1}$
Compressibility	C	$C = 1 / B$	Pa^{-1} = m^2 N^{-1}	[L] [T]^2[M]$^{-1}$

Deformation of Solids

Friction

Friction is a complicated phenomenon. Macroscopically the friction force F between the interface of two materials can be modelled as proportional to the reaction force R at a point of contact between two interfaces, through a dimensionless coefficient of friction μ_f which depends on the pair of materials:

$$F = \mu_f R.$$

This can be applied to static friction (friction preventing two stationary objects from slipping on their own), kinetic friction (friction between two objects scraping/sliding past each other), or rolling (frictional force which prevents slipping but causes a torque to exert on a round object). Surprisingly, the friction force does not depend on the surface area of common contact.

Stress and Strain

The stress-strain constitutive relation for linear materials is commonly known as Hooke's law. In its simplest form, the law defines the spring constant (or elasticity constant) k in a scalar equation, stating the tensile/compressive force is proportional to the extended (or contracted) displacement x:

$$F_i = -kx_i$$

meaning the material responds linearly. Equivalently, in terms of the stress σ, Young's modulus E, and strain ε (dimensionless):

$$\sigma = E\varepsilon$$

In general, forces which deform solids can be normal to a surface of the material (normal forces), or tangential (shear forces), this can be described mathematically using the stress tensor:

$$\sigma_{ij} = C_{ijkl}\varepsilon_{kl} \rightleftharpoons \varepsilon_{ij} = S_{ijkl}\sigma_{kl}$$

where C is the elasticity tensor and S is the compliance tensor

Solid-state Deformations

Several classes of deformations in elastic materials are the following:

- *Elastic*: The material recovers its initial shape after deformation.

- *Anelastic*: if the material is close to elastic, but the applied force induces additional time-dependent resistive forces (i.e. depend on rate of change of extension/compression, in addition to the extension/compression). Metals and ceramics have this characteristic, but it is usually negligible, although not so much when heating due to friction occurs (such as vibrations or shear stresses in machines).

- *Viscoelastic*: If the time-dependent resistive contributions are large, and cannot be neglected. Rubbers and plastics have this property, and certainly do not satisfy Hooke's law. In fact, elastic hysteresis occurs.

- *Plastic*: The applied force induces non-recoverable deformations in the material when the stress (or elastic strain) reaches a critical magnitude, called the yield point.

- *Hyperelastic*: The applied force induces displacements in the material following a strain energy density function.

Collisions

The relative speed of separation $v_{\text{separation}}$ of an object A after a collision with another object B is related to the relative speed of approach v_{approach} by the coefficient of restitution, defined by Newton's experimental impact law:

$$e = \frac{|\mathbf{v}|_{\text{separation}}}{|\mathbf{v}|_{\text{approach}}}$$

which depends the materials A and B are made from, since the collision involves interactions at the surfaces of A and B. Usually $0 \le e \le 1$, in which $e = 1$ for completely elastic collisions, and $e = 0$ for completely inelastic collisions. It's possible for $e \ge 1$ to occur – for superelastic (or explosive) collisions.

Deformation of Fluids

The drag equation gives the drag force D on an object of cross-section area A moving through a fluid of density ρ at velocity v (relative to the fluid)

$$D = \frac{1}{2}c_d \rho A v^2$$

where the drag coefficient (dimensionless) c_d depends on the geometry of the object and the drag forces at the interface between the fluid and object.

For a Newtonian fluid of viscosity μ, the shear stress τ is linearly related to the strain rate (transverse flow velocity gradient) $\partial u / \partial y$ (units s^{-1}). In a uniform shear flow:

$$\tau = \mu \frac{\partial u}{\partial y},$$

with $u(y)$ the variation of the flow velocity u in the cross-flow (transverse) direction y. In general, for a Newtonian fluid, the relationship between the elements τ_{ij} of the shear stress tensor and the deformation of the fluid is given by

$$\tau_{ij} = 2\mu\left(e_{ij} - \frac{1}{3}\Delta\delta_{ij}\right) \quad \text{with} \quad e_{ij} = \frac{1}{2}\left(\frac{\partial v_i}{\partial x_j} + \frac{\partial v_j}{\partial x_i}\right) \quad \text{and} \quad \Delta = \sum_k e_{kk} = \text{div } \mathbf{v},$$

where v_i are the components of the flow velocity vector in the corresponding x_i coordinate directions, e_{ij} are the components of the strain rate tensor, Δ is the volumetric strain rate (or dilatation rate) and δ_{ij} is the Kronecker delta.

The *ideal gas law* is a constitutive relation in the sense the pressure p and volume V are related to the temperature T, via the number of moles n of gas:

$$pV = nRT$$

where R is the gas constant (J K^{-1} mol^{-1}).

Electromagnetism

Constitutive Equations in Electromagnetism and Related Areas

In both classical and quantum physics, the precise dynamics of a system form a set of coupled differential equations, which are almost always too complicated to be solved exactly, even at the level of statistical mechanics. In the context of electromagnetism, this remark applies to not only the dynamics of free charges and currents (which enter Maxwell's equations directly), but also the dynamics of bound charges and currents (which enter Maxwell's equations through the constitutive relations). As a result, various approximation schemes are typically used.

For example, in real materials, complex transport equations must be solved to determine the time and spatial response of charges, for example, the Boltzmann equation or the Fokker–Planck equation or the Navier-Stokes equations. For example, see magnetohydrodynamics, fluid dynamics, electrohydrodynamics, superconductivity, plasma modeling. An entire physical apparatus for dealing with these matters has developed. See for example, linear response theory, Green–Kubo relations and Green's function (many-body theory).

These complex theories provide detailed formulas for the constitutive relations describing the electrical response of various materials, such as permittivities, permeabilities, conductivities and so forth.

It is necessary to specify the relations between displacement field D and E, and the magnetic H-field H and B, before doing calculations in electromagnetism (i.e. applying Maxwell's macroscopic equations). These equations specify the response of bound charge and current to the applied fields and are called constitutive relations.

Determining the constitutive relationship between the auxiliary fields D and H and the E and B fields starts with the definition of the auxiliary fields themselves:

$$\mathbf{D}(\mathbf{r},t) = \varepsilon_0 \mathbf{E}(\mathbf{r},t) + \mathbf{P}(\mathbf{r},t)$$

$$\mathbf{H}(\mathbf{r},t) = \frac{1}{\mu_0} \mathbf{B}(\mathbf{r},t) - \mathbf{M}(\mathbf{r},t),$$

where P is the polarization field and M is the magnetization field which are defined in terms of microscopic bound charges and bound current respectively. Before getting to how to calculate M and P it is useful to examine the following special cases.

Without Magnetic or Dielectric Materials

In the absence of magnetic or dielectric materials, the constitutive relations are simple:

$$\mathbf{D} = \varepsilon_0 \mathbf{E}, \quad \mathbf{H} = \mathbf{B} / \mu_0$$

where ε_0 and μ_0 are two universal constants, called the permittivity of free space and permeability of free space, respectively.

Isotropic Linear Materials

In an (isotropic) linear material, where P is proportional to E, and M is proportional to B, the constitutive relations are also straightforward. In terms of the polarization P and the magnetization M they are:

$$\mathbf{P} = \varepsilon_0 \chi_e \mathbf{E}, \quad \mathbf{M} = \chi_m \mathbf{H},$$

where χ_e and χ_m are the electric and magnetic susceptibilities of a given material respectively. In terms of D and H the constitutive relations are:

$$\mathbf{D} = \varepsilon \mathbf{E}, \quad \mathbf{H} = \mathbf{B} / \mu,$$

where ε and μ are constants (which depend on the material), called the permittivity and permeability, respectively, of the material. These are related to the susceptibilities by:

$$\varepsilon / \varepsilon_0 = \varepsilon_r = (\chi_e + 1), \quad \mu / \mu_0 = \mu_r = (\chi_m + 1)$$

General Case

For real-world materials, the constitutive relations are not linear, except approximately. Calculating the constitutive relations from first principles involves determining how P and M are created from a given E and B. These relations may be empirical (based directly upon measurements), or theoretical (based upon statistical mechanics, transport theory or other tools of condensed matter physics). The detail employed may be macroscopic or microscopic, depending upon the level necessary to the problem under scrutiny.

In general, the constitutive relations can usually still be written:

$$\mathbf{D} = \varepsilon \mathbf{E}, \quad \mathbf{H} = \mu^{-1} \mathbf{B}$$

but ε and μ are not, in general, simple constants, but rather functions of E, B, position and time, and tensorial in nature. Examples are:

- *Dispersion and absorption* where ε and μ are functions of frequency. (Causality does not permit materials to be nondispersive; see, for example, Kramers–Kronig relations). Neither do the fields need to be in phase which leads to ε and μ being complex. This also leads to absorption.

- *Nonlinearity* where ε and μ are functions of E and B.

- *Anisotropy* (such as *birefringence* or *dichroism*) which occurs when ε and μ are second-rank tensors,

$$D_i = \sum_j \varepsilon_{ij} E_j \quad B_i = \sum_j \mu_{ij} H_j.$$

- Dependence of P and M on E and B at other locations and times. This could be due to *spatial inhomogeneity*; for example in a domained structure, heterostructure or a liquid crystal, or most commonly in the situation where there are simply multiple materials occupying different regions of space. Or it could be due to a time varying medium or due to hysteresis. In such cases P and M can be calculated as:

$$\mathbf{P}(\mathbf{r},t) = \varepsilon_0 \int d^3\mathbf{r}' dt' \ \hat{\chi}_e(\mathbf{r},\mathbf{r}',t,t';\mathbf{E}) \mathbf{E}(\mathbf{r}',t')$$

$$\mathbf{M}(\mathbf{r},t) = \frac{1}{\mu_0} \int d^3\mathbf{r}' dt' \ \hat{\chi}_m(\mathbf{r},\mathbf{r}',t,t';\mathbf{B}) \mathbf{B}(\mathbf{r}',t'),$$

in which the permittivity and permeability functions are replaced by integrals over the more general electric and magnetic susceptibilities. In homogenous materials, dependence on other locations is known as spatial dispersion.

As a variation of these examples, in general materials are bianisotropic where D and B depend on both E and H, through the additional *coupling constants* ξ and ζ:

$$\mathbf{D} = \varepsilon\mathbf{E} + \xi\mathbf{H}, \quad \mathbf{B} = \mu\mathbf{H} + \zeta\mathbf{E}.$$

In practice, some materials properties have a negligible impact in particular circumstances, permitting neglect of small effects. For example: optical nonlinearities can be neglected for low field strengths; material dispersion is unimportant when frequency is limited to a narrow bandwidth; material absorption can be neglected for wavelengths for which a material is transparent; and metals with finite conductivity often are approximated at microwave or longer wavelengths as perfect metals with infinite conductivity (forming hard barriers with zero skin depth of field penetration).

Some man-made materials such as metamaterials and photonic crystals are designed to have customized permittivity and permeability.

Calculation of Constitutive Relations

The theoretical calculation of a material's constitutive equations is a common, important, and sometimes difficult task in theoretical condensed-matter physics and materials science. In general, the constitutive equations are theoretically determined by calculating how a molecule responds to the local fields through the Lorentz force. Other forces may need to be modeled as well such as lattice vibrations in crystals or bond forces. Including all of the forces leads to changes in the molecule which are used to calculate P and M as a function of the local fields.

The local fields differ from the applied fields due to the fields produced by the polarization and magnetization of nearby material; an effect which also needs to be modeled. Further, real materials are not continuous media; the local fields of real materials vary wildly on the atomic scale. The fields need to be averaged over a suitable volume to form a continuum approximation.

These continuum approximations often require some type of quantum mechanical analysis such as quantum field theory as applied to condensed matter physics. See, for example, density functional theory, Green–Kubo relations and Green's function.

A different set of *homogenization methods* (evolving from a tradition in treating materials such as conglomerates and laminates) are based upon approximation of an inhomogeneous material by a homogeneous *effective medium* (valid for excitations with wavelengths much larger than the scale of the inhomogeneity).

The theoretical modeling of the continuum-approximation properties of many real materials often rely upon experimental measurement as well. For example, ε of an insulator at low frequencies can be measured by making it into a parallel-plate capacitor, and ε at optical-light frequencies is often measured by ellipsometry.

Thermoelectric and Electromagnetic Properties of Matter

These constitutive equations are often used in crystallography, a field of solid-state physics.

Electromagnetic properties of solids			
Property/effect	**Stimuli/response parameters of system**	**Constitutive tensor of system**	**Equation**
Hall effect	E = electric field strength (N C^{-1}) J = electric current density (A m^{-2}) H = magnetic field intensity (A m^{-1})	ρ = electrical resistivity (Ω m)	$E_k = \rho_{kij} J_i H_j$
Direct Piezoelectric Effect	σ = Stress (Pa) P = (dielectric) polarization (C m^{-2})	d = direct piezoelectric coefficient (K^{-1})	$P_i = d_{ijk}\sigma_{jk}$
Converse Piezoelectric Effect	ε = Strain (dimensionless) E = electric field strength (N C^{-1})	d = direct piezoelectric coefficient (K^{-1})	$\varepsilon_{ij} = d_{ijk} E_k$
Piezomagnetic effect	σ = Stress (Pa) M = magnetization (A m^{-1})	q = piezomagnetic coefficient (K^{-1})	$M_i = q_{ijk}\sigma_{jk}$

Thermoelectric properties of solids			
Property/effect	**Stimuli/response parameters of system**	**Constitutive tensor of system**	**Equation**
Pyroelectricity	P = (dielectric) polarization (C m^{-2}) T = temperature (K)	p = pyroelectric coefficient (C m^{-2} K^{-1})	$\Delta P_j = p_j \Delta T$
Electrocaloric effect	S = entropy (J K^{-1}) E = electric field strength (N C^{-1})	p = pyroelectric coefficient (C m^{-2} K^{-1})	$\Delta S = p_i \Delta E_i$
Seebeck effect	E = electric field strength (N C^{-1} = V m^{-1}) T = temperature (K) x = displacement (m)	β = thermopower (V K^{-1})	$E_i = -\beta_{ij}\dfrac{\partial T}{\partial x_j}$
Peltier effect	E = electric field strength (N C^{-1}) J = electric current density (A m^{-2}) q = heat flux (W m^{-2})	Π = Peltier coefficient (W A^{-1})	$q_j = \Pi_{ji} J_i$

Photonics

Refractive index

The (absolute) refractive index of a medium n (dimensionless) is an inherently important property of geometric and physical optics defined as the ratio of the luminal speed in vacuum c_0 to that in the medium c:

$$n = \frac{c_0}{c} = \sqrt{\frac{\varepsilon\mu}{\varepsilon_0\mu_0}} = \sqrt{\varepsilon_r\mu_r}$$

where ε is the permittivity and ε_r the relative permittivity of the medium, likewise μ is the permeability and μ_r are the relative permmeability of the medium. The vacuum permittivity is ε_0 and vacuum permeability is μ_0. In general, n (also ε_r) are complex numbers.

The relative refractive index is defined as the ratio of the two refractive indices. Absolute is for on material, relative applies to every possible pair of interfaces;

$$n_{AB} = \frac{n_A}{n_B}$$

Speed of light in matter

As a consequence of the definition, the speed of light in matter is

$$c = 1/\sqrt{\varepsilon\mu}$$

for special case of vacuum; $\varepsilon = \varepsilon_0$ and $\mu = \mu_0$,

$$c_0 = 1/\sqrt{\varepsilon_0\mu_0}$$

Piezooptic effect

The piezooptic effect relates the stresses in solids σ to the dielectric impermeability a, which are coupled by a fourth-rank tensor called the piezooptic coefficient Π (units K^{-1}):

$$a_{ij} = \Pi_{ijpq}\sigma_{pq}$$

Transport Phenomena

Definitions

Definitions (thermal properties of matter)				
Quantity (Common Name/s)	**(Common) Symbol/s**	**Defining Equation**	**SI Units**	**Dimension**
General heat capacity	C = heat capacity of substance	$q = CT$	J K^{-1}	$[M][L]^2[T]^{-2}[\Theta]^{-1}$
Linear thermal expansion	L = length of material (m) α = coefficient linear thermal expansion (dimensionless) ε = strain tensor (dimensionless)	$\partial L/\partial T = \alpha L$ $\varepsilon_{ij} = \alpha_{ij}\Delta T$	K^{-1}	$[\Theta]^{-1}$
Volumetric thermal expansion	β, γ V = volume of object (m³) p = constant pressure of surroundings	$(\partial V/\partial T)_p = \gamma V$	K^{-1}	$[\Theta]^{-1}$

Thermal conductivity	κ, K, λ, **A** = surface cross section of material (m²) P = thermal current/power through material (W) ∇T = temperature gradient in material (K m⁻¹)	$\lambda = -P/(\mathbf{A}\cdot\nabla T)$	W m⁻¹ K⁻¹	[M][L][T]⁻³[Θ]⁻¹
Thermal conductance	U	$U = \lambda/\delta x$	W m⁻² K⁻¹	[M][T]⁻³[Θ]⁻¹
Thermal resistance	R Δx = displacement of heat transfer (m)	$R = 1/U = \Delta x/\lambda$	m² K W⁻¹	[M]⁻¹[L][T]³[Θ]

Definitions (Electrical/magnetic properties of matter)				
Quantity (Common Name/s)	**(Common) Symbol/s**	**Defining Equation**	**SI Units**	**Dimension**
Electrical resistance	R	$R = V >$	Ω = V A⁻¹ = J s C⁻²	[M] [L]² [T]⁻³ [I]⁻²
Resistivity	ρ	$\rho = RA/l$	Ω m	[M]² [L]² [T]⁻³ [I]⁻²
Resistivity temperature coefficient, linear temperature dependence	α	$\rho - \rho_0 = \rho_0 \alpha(T-T_0)$	K⁻¹	[Θ]⁻¹
Electrical conductance	G	$G = 1 >$	S = Ω⁻¹	[T]³ [I]² [M]⁻¹ [L]⁻²
Electrical conductivity	σ	$\sigma = 1/\rho$	Ω⁻¹ m⁻¹	[I]² [T]³ [M]⁻² [L]⁻²
Magnetic reluctance	$R, R_m, \mathcal{R}$	$R_m = \mathcal{M}/\Phi_B$	A Wb⁻¹ = H⁻¹	[M]⁻¹[L]⁻²[T]²
Magnetic permeance	$P, P_m, \Lambda, \mathcal{P}$	$\Lambda = 1/R_m$	Wb A⁻¹ = H	[M][L]²[T]⁻²

Definitive Laws

There are several laws which describe the transport of matter, or properties of it, in an almost identical way. In every case, in words they read:

Flux (density) is proportional to a gradient, the constant of proportionality is the characteristic of the material.

In general the constant must be replaced by a 2nd rank tensor, to account for directional dependences of the material.

Property/effect	**Nomenclature**	**Equation**
Fick's law of diffusion, defines diffusion coefficient D	D = mass diffusion coefficient (m² s⁻¹) J = diffusion flux of substance (mol m⁻² s⁻¹) $\partial C/\partial x$ = (1d)concentration gradient of substance (mol dm⁻⁴)	$J_i = -D_{ij}\dfrac{\partial C}{\partial x_j}$

Darcy's law for fluid flow in porous media, defines permeability κ	κ = permeability of medium (m²) μ = fluid viscosity (Pa s) q = discharge flux of substance (m s⁻¹) ∂P/∂x = (1d) pressure gradient of system (Pa m⁻¹)	$q_j = -\dfrac{\kappa}{\mu}\dfrac{\partial P}{\partial x_j}$
Ohm's law of electric conduction, defines electric conductivity (and hence resistivity and resistance)	V = potential difference in material (V) I = electric current through material (A) R = resistance of material (Ω) ∂V/∂x = potential gradient (electric field) through material (V m⁻¹) J = electric current density through material (A m⁻²) σ = electric conductivity of material (Ω⁻¹ m⁻¹) ρ = electrical resistivity of material (Ω m)	Simplist form is: $V = IR$ More general forms are: $\dfrac{\partial V}{\partial x_j} = \rho_{ji} J_i \rightleftharpoons J_i = \sigma_{ij}\dfrac{\partial V}{\partial x_j}$
Fourier's law of thermal conduction, defines thermal conductivity λ	λ = thermal conductivity of material (W m⁻¹ K⁻¹) q = heat flux through material (W m⁻²) ∂T/∂x = temperature gradient in material (K m⁻¹)	$q_i = -\lambda_{ij}\dfrac{\partial T}{\partial x_j}$
Stefan–Boltzmann law of black-body radiation, defines emmisivity ε	I = radiant intensity (W m⁻²) σ = Stefan–Boltzmann constant (W m⁻² K⁻⁴) T_{sys} = temperature of radiating system (K) T_{ext} = temperature of external surroundings (K) ε = emissivity (dimensionless)	For a single radiator: $I = \varepsilon\sigma T^4$ For a temperature difference: $I = \varepsilon\sigma(T_{ext}^4 - T_{sys}^4)$ $0 \leq \varepsilon \leq 1$ ε = 0 for perfect reflector ε = 1 for perfect absorber (true black body)

3D Constitutive Equations

In this section, we are going to develop the 3D constitutive equations. We will start with the the generalized Hooke's law for a material, that is, material is generally anisotropic in nature. Finally, we will derive the constitutive equation for isotropic material, with which the readers are very familiar. The journey for constitutive equation from anisotropic to isotropic material is very interesting and will use most of the concepts.

The generalized Hooke's law for a material is given as

$$\sigma_{ij} = C_{ijkl}\varepsilon_{kl} \quad i,j,k,l = 1,2,3 \tag{1}$$

where, σ_{ij} is a second order tensor known as stress tensor and its individual elements are the stress components. ε_{ij} is another second order tensor known as strain tensor and its individual elements are the strain components. C_{ijkl} is a fourth order tensor known as stiffness tensor. In the remaining section we will call it as stiffness matrix, as popularly known. The individual elements of this tensor are the stiffness coefficients for this linear stress-strain relationship. Thus, stress and strain tensor has $(3 \times 3 =)$ 9 components each and the stiffness tensor has $((3)^4 =)$ 81 independent elements. The individual elements =) 81 are referred by various names as elastic constants, moduli and stiffness coefficients. The reduction in the number of these elastic constants can be sought with the following symmetries.

Stress Symmetry

The stress components are symmetric under this symmetry condition, that is, $\sigma_{ij} = \sigma_{jl}$. Thus, there are six independent stress components. Hence, from Equation. (1) we write

$$\sigma_{ji} = C_{jikl} \varepsilon_{kl} \tag{2}$$

Subtracting Equation (2) from Equation (1) leads to the following equation

$$0 = \left(C_{ijkl} - C_{jikl} \right) \varepsilon_{kl} \Rightarrow C_{ijkl} = C_{jikl} \tag{3}$$

There are six independent ways to express i and j taken together and still nine independent ways to express k and l taken together. Thus, with stress symmetry the number of independent elastic constants reduce to $(6 \times 9 =)$ 54 from 81.

Strain Symmetry

The strain components are symmetric under this symmetry condition, that is, $\varepsilon_{ij} = \varepsilon_{ji}$. Hence, from Equation (1) we write

$$\sigma_{ij} = C_{ijkl} \varepsilon_{ik}$$

Subtracting Equation (3) from Equation (2) we get the following equation

$$0 = \left(C_{ijkl} - C_{ijlk} \right) \varepsilon_{kl} \Rightarrow C_{ijkl} = C_{ijlk} \tag{4}$$

It can be seen from Equation (4) that there are six independent ways of expressing i and j taken together when k and l are fixed. Similarly, there are six independent ways of expressing k and l taken together when i and j are fixed in Equation (4). Thus, there are $6 \times 6 = 36$ independent constants for this linear elastic material with stress and strain symmetry.

With this reduced stress and strain components and reduced number of stiffness coefficients, we can write Hooke's law in a contracted form as

$$\sigma_i = C_{ij} \varepsilon_j \quad (i, j = 1, 2, \cdots, 6) \tag{5}$$

where

$$\sigma_1 = \sigma_{11} \qquad \varepsilon_1 = \varepsilon_{11}$$

$$\sigma_2 = \sigma_{22} \qquad \varepsilon_2 = \varepsilon_{22}$$

$$\sigma_3 = \sigma_{33} \qquad \varepsilon_3 = \varepsilon_{33}$$

$$\sigma_4 = \sigma_{23} \qquad \varepsilon_4 = 2\varepsilon_{23}$$

$$\sigma_5 = \sigma_{13} \qquad \varepsilon_5 = \varepsilon_{13}$$

$$\sigma_6 = \sigma_{12} \qquad \varepsilon_6 = 2\varepsilon_{12} \qquad (6)$$

Note: The shear strains are the engineering shear strains.

For Equation (5) to be solvable for strains in terms of stresses, the determinant of the stiffness matrix must be nonzero, that is $\left| C_{ij} \right| \neq 0$.

The number of independent elastic constants can be reduced further, if there exists strain energy density function W, given as below.

Strain Energy Density Function (*W*)

The strain energy density function *W* is given as

$$W = \frac{1}{2} C_{ji} \varepsilon_j \varepsilon_i \qquad (7)$$

with the property that

$$\sigma_i = \frac{\partial W}{\partial \varepsilon_i} \qquad (8)$$

It is seen that W is a quadratic function of strain. A material with the existence of W with property in Equation (8) is called as Hyperelastic Material.

The W can also be written as

$$W = \frac{1}{2} C_{ji} \varepsilon_j \varepsilon_i \qquad (9)$$

Subtracting Equation (9) from Equation (7) we get

$$0 = \left(C_{ij} - C_{ji} \right) \varepsilon_i \varepsilon_j \qquad (10)$$

which leads to the identity $\left(C_{ij} = C_{ji} \right)$. Thus, the stiffness matrix is symmetric. This symmetric matrix has 21 independent elastic constants. The stiffness matrix is given as follows:

$$C_{ij} = \begin{bmatrix} C_{11} & C_{12} & C_{13} & C_{14} & C_{15} & C_{16} \\ & C_{22} & C_{23} & C_{24} & C_{25} & C_{26} \\ & & C_{33} & C_{34} & C_{35} & C_{36} \\ & & & C_{44} & C_{45} & C_{46} \\ & Symmetric & & & C_{55} & C_{56} \\ & & & & & C_{66} \end{bmatrix} \quad (11)$$

The existence of the function W is based upon the first and second law of thermodynamics. Further, it should be noted that this function is positive definite. Also, the function W is an invariant (An invariant is a quantity which is independent of change of reference).

The material with 21 independent elastic constants is called Anisotropic or Aelotropic Material. Further reduction in the number of independent elastic constants can be obtained with the use of planes of material symmetry as follows.

Material Symmetry

It should be recalled that both the stress and strain tensor follow the transformation rule and so does the stiffness tensor. The transformation rule for these quantities (as given in Equation (1)) is known as follows

$$\sigma'_{ij} = a_{ki} a_{ij} \sigma_{kl}$$
$$\varepsilon'_{ij} = a_{ki} a_{ij} \varepsilon_{kl}$$
$$C'_{ijkl} = a_{pi} a_{qi} a_{rk} a_{sl} C_{pqr} \quad (12)$$

where α_{ij} are the direction cosines from i to j coordinate system. The prime indicates the quantity in new coordinate system.

When the function W given in Equation (9) is expanded using the contracted notations for strains and elastic constants given in Equation (11) W has the following form:

$$W = \frac{1}{2} \begin{bmatrix} C_{11}\varepsilon_1^2 + 2C_{12}\varepsilon_1\varepsilon_2 + 2C_{13}\varepsilon_1\varepsilon_3 + 2C_{14}\varepsilon_1\varepsilon_4 + 2C_{15}\varepsilon_1\varepsilon_5 + 2C_{16}\varepsilon_1\varepsilon_6 + \\ C_{22}\varepsilon_2^2 + 2C_{23}\varepsilon_2\varepsilon_3 + 2C_{24}\varepsilon_2\varepsilon_4 + 2C_{25}\varepsilon_2\varepsilon_5 + 2C_{26}\varepsilon_2\varepsilon_6 + \\ C_{33}\varepsilon_3^2 + 2C_{34}\varepsilon_3\varepsilon_4 + 2C_{35}\varepsilon_3\varepsilon_5 + 2C_{36}\varepsilon_3\varepsilon_6 + \\ C_{44}\varepsilon_4^2 + 2C_{45}\varepsilon_4\varepsilon_5 + 2C_{46}\varepsilon_4\varepsilon_6 + \\ C_{55}\varepsilon_5^2 + 2C_{56}\varepsilon_5\varepsilon_6 + \\ C_{66}\varepsilon_6^2 \end{bmatrix} \quad (13)$$

Thus, from Equation (13) it can be said that the function W has the following form in terms of strain components:

$$W = W \begin{bmatrix} \varepsilon_1^2, \varepsilon_2^2, \varepsilon_3^2, \varepsilon_4^2, \varepsilon_5^2, \varepsilon_6^2, \\ \varepsilon_1\varepsilon_2, \varepsilon_1\varepsilon_3, \varepsilon_1\varepsilon_4, \varepsilon_1\varepsilon_5, \varepsilon_1\varepsilon_6, \\ \varepsilon_2\varepsilon_3, \varepsilon_2\varepsilon_4, \varepsilon_2\varepsilon_5, \varepsilon_2\varepsilon_6, \\ \varepsilon_3\varepsilon_4, \varepsilon_3\varepsilon_5, \varepsilon_3\varepsilon_6, \\ \varepsilon_4\varepsilon_5, \varepsilon_4\varepsilon_6, \\ \varepsilon_5\varepsilon_6, \end{bmatrix} \tag{14}$$

With these concepts we proceed to consider the planes of material symmetry. The planes of the material, also called elastic symmetry are due to the symmetry of the structure of anisotropic body. In the following, we consider some special cases of material symmetry.

Symmetry with Respect to a Plane

Let us assume that the anisotropic material has only one plane of material symmetry. A material with one plane of material symmetry is called Monoclinic Material.

Let us consider the x_1-x_2 ($x_3 = 0$) plane as the plane of material symmetry. This is shown in Figure. This symmetry can be formulated with the change of axes as follows:

$$x_1' = x_1, \quad x_2' = x_2, \quad x_3' = -x_3 \tag{15}$$

With this change of axes,

$$\alpha_{ij} = \frac{\partial x_j'}{\partial x_j} \quad and \quad \frac{\partial x_i'}{\partial x_j} = \delta_{ij} \ \ for \ j = 1, 2 \ and \ \frac{\partial x_i'}{\partial x_3} = -\delta_{i3} \tag{16}$$

This gives us along with the use of the second of Equation (12)

$$\varepsilon_{11}' = \varepsilon_{11}, \varepsilon_{22}' = \varepsilon_{22}, \varepsilon_{33}' = \varepsilon_{33}, \varepsilon_{23}' = -\varepsilon_{23}, \varepsilon_{13}' = -\varepsilon_{13}, \varepsilon_{12}' = \varepsilon_{12} \tag{17}$$

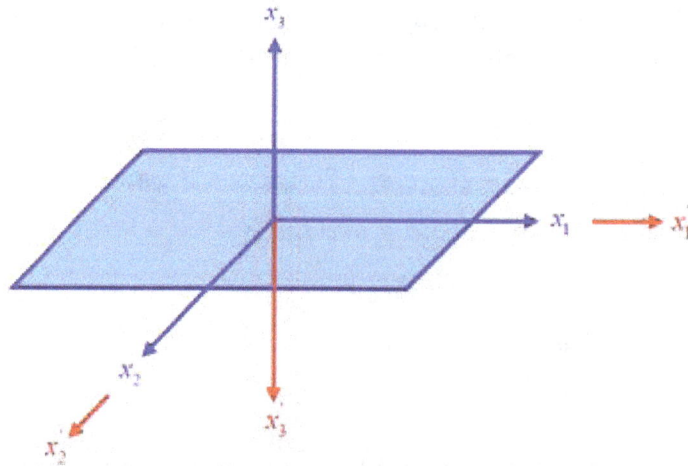

Material symmetry about *x1-x2* plane

First Approach: Invariance Approach

Now, the function W can be expressed in terms of the strain components ε'_{ij}. If W is to be invariant, then it must be of the form

$$W = W\left[\varepsilon_1^2, \varepsilon_2^2, \varepsilon_3^2, \varepsilon_4^2, \varepsilon_5^2, \varepsilon_6^2, \varepsilon_1\varepsilon_2, \varepsilon_1\varepsilon_3, \varepsilon_1\varepsilon_6, \varepsilon_2\varepsilon_3, \varepsilon_2\varepsilon_6, \varepsilon_3\varepsilon_6, \varepsilon_4\varepsilon_5,\right] \tag{18}$$

Comparing this with Equation (13) it is easy to conclude that

$$C_{14} = C_{15} = C_{24} = C_{25} = C_{34} = C_{35} = C_{46} = C_{56} = 0 \tag{19}$$

Thus, for the monoclinic materials the number of independent constants are 13. With this reduction of number of independent elastic constants the stiffness matrix is given as

$$C_{ij} = \begin{bmatrix} C_{11} & C_{12} & C_{13} & 0 & 0 & C_{16} \\ & C_{22} & C_{23} & 0 & 0 & C_{26} \\ & & C_{33} & 0 & 0 & C_{36} \\ & & & C_{44} & C_{45} & 0 \\ & Symmetric & & & C_{55} & 0 \\ & & & & & C_{66} \end{bmatrix} \tag{20}$$

Second Approach: Stress Strain Equivalence Approach

The same reduction of number of elastic constants can be derived from the stress strain equivalence approach. From Equation (12) and Equation (16) we have

$$\sigma'_{11} = \sigma_{11}, \sigma'_{22} = \sigma_{22}, \sigma'_{33} = \sigma_{33}, \sigma'_{23} = -\sigma_{23}, \sigma'_{13} = -\sigma_{13}, \sigma'_{12} = \sigma_{12} \tag{21}$$

The same can be seen from the stresses on a cube inside such a body with the coordinate systems shown in Figure. Figure (a) shows the stresses on a cube with the coordinate system x_1, x_2, x_3 and Figure (b) shows stresses on the same cube with the coordinate system x'_1, x'_2, x'_3.. Comparing the stresses we get the relation as in Equation (21).

Now using the stiffness matrix as given in Equation (11), strain term relations as given in Equation (17) and comparing the stress terms in Equation (21) as follows:

$$\sigma_{11} = \sigma'_{11}$$
$$C_{11}\varepsilon_1 + C_{12}\varepsilon_2 + C_{13}\varepsilon_3 + C_{14}\varepsilon_4 + C_{15}\varepsilon_5 + C_{16}\varepsilon_6 = C'_{11}\varepsilon'_1 + C'_{12}\varepsilon'_2 + C_{13}\varepsilon'_3 + C'_{14}\varepsilon'_4 + C'_{15}\varepsilon'_5 + C'_{16}\varepsilon'_6$$

Using the relations from Equation (3.17), the above equations reduce to

$$C_{14}\varepsilon_4 + C_{15}\varepsilon_5 = C'_{14}\varepsilon'_4 + C'_{15}\varepsilon'_5$$

Noting that $C_{ij} = C'_{ij}$, this holds true only when $C_{14} = C_{15} = 0$

Similarly,

$$\sigma_{22} = \sigma'_{22} \ gives \ C_{24} = C_{25} = 0$$
$$\sigma_{33} = \sigma'_{33} \ gives \ C_{34} = C_{35} = 0$$
$$\sigma_{23} = \sigma'_{23} \ gives \ C_{46} = 0$$
$$\sigma_{13} = \sigma'_{13} \ gives \ C_{56} = 0$$

This gives us the C_{jk} matrix as in Equation (20).

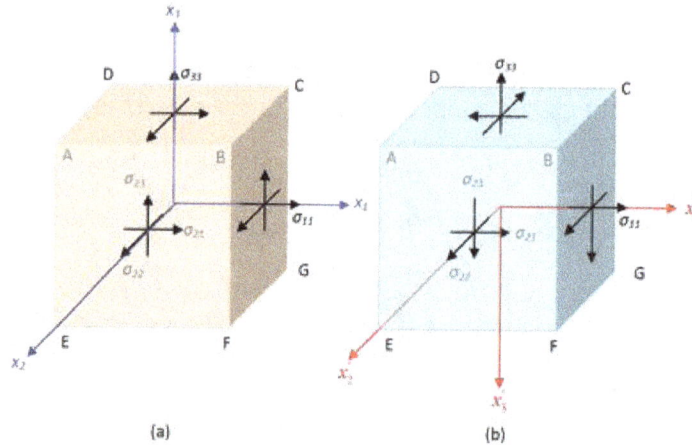

State of stress (a) in x_1, x_2, x_3 system
(b) with x_1-x_3 plane of symmetry

Symmetry with Respect to Two Orthogonal Planes

Let us assume that the material under consideration has one more plane, say x_2-x_3 is plane of material symmetry along with x_1-x_2 as in (A). These two planes are orthogonal to each other. This transformation is shown in Figure.

This can be mathematically formulated by the change of axes as

$$x'_1 = -x_1, x'_2 = x_2, x'_3 = -x_3 \tag{22}$$

And

$$\alpha_{ij} = \frac{\partial x'_i}{\partial x_j} \ and \ \frac{\partial x'_i}{\partial x_j} = -\delta_{ij} \ for \ j = 1,3 \ and \ \frac{\partial x'_i}{\partial x_2} = \delta_{i2} \tag{23}$$

This gives us the required strain relations as (from Equation (12)).

$$\varepsilon'_{11} = \varepsilon_{11}, \varepsilon'_{22} = \varepsilon_{22}, \varepsilon'_{33} = \varepsilon_{33}, \varepsilon'_{23} = -\varepsilon_{23}, \varepsilon'_{13} = \varepsilon_{13}, \varepsilon'_{12} = -\varepsilon_{12}$$

or using contracted notations, we can write,

$$\varepsilon'_1 = \varepsilon_1, \varepsilon'_2 = \varepsilon_2, \varepsilon'_3 = \varepsilon_3, \varepsilon'_4 = -\varepsilon_4, \varepsilon'_5 = \varepsilon_5, \varepsilon'_6 = -\varepsilon_6 \tag{24}$$

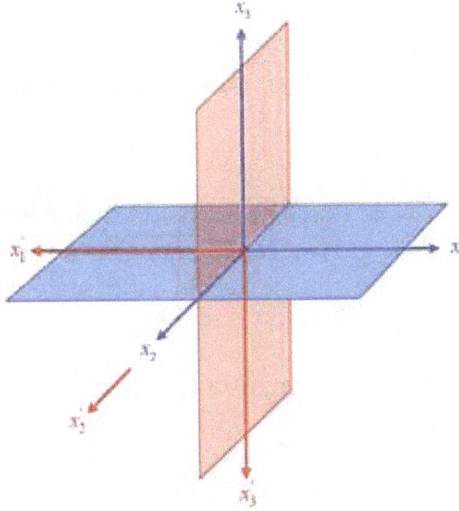

Material symmetry about x_1-x_2 and x_2-x_3 planes

First Approach: Invariance Approach

We can get the function W simply by substituting ε'_{ij} in place of ε_{ij} and using contracted notations for the strains in Equation (18). Noting that W is invariant, its form in Equation (18) must now be restricted to functional form

$$W = W\left[\varepsilon_1^2, \varepsilon_2^2, \varepsilon_3^2, \varepsilon_4^2, \varepsilon_5^2, \varepsilon_6^2, \varepsilon_1\varepsilon_2, \varepsilon_1\varepsilon_3, \varepsilon_2\varepsilon_3\right] \qquad (25)$$

From this it is easy to see that

$$C_{16} = C_{26} = C_{36} = C_{45} = 0$$

Thus, the number of independent constants reduces to 9. The resulting stiffness matrix is given as

$$C_{ij} = \begin{bmatrix} C_{11} & C_{12} & C_{13} & 0 & 0 & 0 \\ & C_{22} & C_{23} & 0 & 0 & 0 \\ & & C_{33} & 0 & 0 & 0 \\ & & & C_{44} & 0 & 0 \\ & \text{Symmetric} & & & C_{55} & 0 \\ & & & & & C_{66} \end{bmatrix} \qquad (26)$$

When a material has (any) two orthogonal planes as planes of material symmetry then that material is known as Orthotropic Material. It is easy to see that when two orthogonal planes are planes of material symmetry, the third mutually orthogonal plane is also plane of material symmetry and Equation (26) holds true for this case also.

Note: Unidirectional fibrous composites are an example of orthotropic materials.

Second Approach: Stress Strain Equivalence Approach

The same reduction of number of elastic constants can be derived from the stress strain equivalence approach. From the first of Equation (12) and Equation (23) we have

$$\sigma_{11}' = \sigma_{11}, \sigma_{22}' = \sigma_{22}, \sigma_{33}' = \sigma_{33}, \sigma_{23}' = -\sigma_{23}, \sigma_{13}' = \sigma_{13}, \sigma_{12}' = -\sigma_{12} \quad (27)$$

The same can be seen from the stresses on a cube inside such a body with the coordinate systems shown in Figure. Figure (a) shows the stresses on a cube with the coordinate system x_1, x_2, x_3 and Figure (b) shows stresses on the same cube with the coordinate system x_1', x_2', x_3'. Comparing the stresses we get the relation as in Equation (27).

Now using the stiffness matrix given in Equation (20) and comparing the stress equivalence of Equation (27) we get the following:

$$\sigma_{11} = \sigma_{11}'$$
$$C_{11}\varepsilon_1 + C_{12}\varepsilon_2 + C_{13}\varepsilon_3 + C_{16}\varepsilon_6 = C_{11}'\varepsilon_1' + C_{12}'\varepsilon_2' + C_{13}'\varepsilon_3' + C_{16}'\varepsilon_6'$$

This holds true when $C_{16} = 0$. Similarly,

$$\sigma_{22} = \sigma_{22}' \; gives \; C_{26} = 0$$
$$\sigma_{33} = \sigma_{33}' \; gives \; C_{36} = 0$$
$$\sigma_{23} = -\sigma_{23}' \left(or \; \sigma_{13} = \sigma_{13}' \right) gives \; C_{45} = 0$$

This gives us the C_{ij} matrix as in Equation (26).

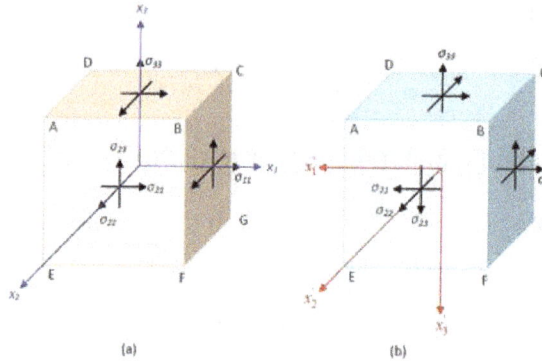

State of stress (a) in x_1, x_2, x_3 system (b) with x_1-x_2 and x_2-x_3 planes of symmetry

Alternately, if we consider x_1-x_3 as the second plane of material symmetry along with x_1-x_2 as shown in Figure, then

$$x_1' = x_1, x_2' = -x_2, x_3' = -x_3 \quad (28)$$

and

$$a_{ij} = \frac{\partial x_i'}{\partial x_j} \; and \; \frac{\partial x_i'}{\partial x_j} = -\delta_{ij} \; for \; j = 2, 3 \; and \; \frac{\partial x_i'}{\partial x_1} = \delta_{i1} \quad (29)$$

This gives us the required strain relations as (from Equation (12))

$$\varepsilon'_{11} = \varepsilon_{11}, \varepsilon'_{22} = \varepsilon_{22}, \varepsilon'_{33} = \varepsilon_{33}, \varepsilon'_{23} = \varepsilon_{23}, \varepsilon'_{13} = -\varepsilon_{13}, \varepsilon'_{12} = -\varepsilon_{12}$$

or in contracted notations, we write

$$\varepsilon'_1 = \varepsilon_1, \varepsilon'_2 = \varepsilon_2, \varepsilon'_3 = \varepsilon_3, \varepsilon'_4 = \varepsilon_4, \varepsilon'_5 = -\varepsilon_5, \varepsilon'_6 = -\varepsilon_6$$

Substituting these in Equation (18) the function W reduces again to the form given in Equation (25) for W to be invariant. Finally, we get the reduced stiffness matrix as given in Equation (26).

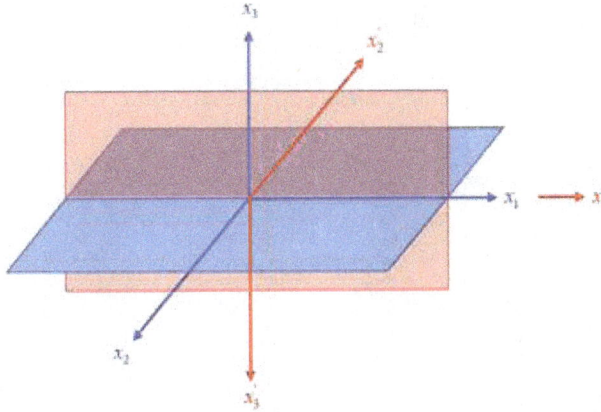

Material symmetry about x_1-x_2 and x_1-x_3 planes

The stress transformations for this coordinate transformations are (from the first of Equation (12) and Equation (29))

$$\sigma'_{11} = \sigma_{11}, \sigma'_{22} = \sigma_{22}, \sigma'_{33} = \sigma_{33}, \sigma'_{23} = \sigma_{23}, \sigma'_{13} = -\sigma_{13}, \sigma'_{12} = -\sigma_{12}$$

The same can be seen from the stresses shown on the same cube in x_1, x_2, x_3 and x'_1, x'_2, x'_3 coordinate systems in Figure (a) and (b), respectively. The comparison of the stress terms leads to the stiffness matrix as given in Equation (26).

Note: It is clear that if any two orthogonal planes are planes of material symmetry the third mutually orthogonal plane has to be plane to material symmetry. We have got the same stiffness matrix when we considered two sets of orthogonal planes. Further, if we proceed in this way considering three mutually orthogonal planes of symmetry then it is not difficult to see that the stiffness matrix remains the same as in Equation (26).

Transverse Isotropy

First Approach: Invariance Approach

This is obtained from an orthotropic material. Here, we develop the constitutive relation for a material with transverse isotropy in x_2-x_3 plane (this is used in lamina/laminae/laminate modeling). This is obtained with the following form of the change of axes.

$$x_1' = x_1$$
$$x_1' = x_2 \cos \alpha + x_3 \sin \alpha$$
$$x_3' = -x_2 \sin \alpha + x_3 \cos \alpha$$

(30)

Now, we have

$$\frac{\partial x_1'}{\partial x_1} = 1, \frac{\partial x_2'}{\partial x_2} = \frac{\partial x_3'}{\partial x_3} = \cos \alpha, \frac{\partial x_2'}{\partial x_3} = -\frac{\partial x_3'}{\partial x_2} = \sin \alpha,$$

$$\frac{\partial x_1'}{\partial x_2} = \frac{\partial x_1'}{\partial x_3} = \frac{\partial x_2'}{\partial x_1} = \frac{\partial x_3'}{\partial x_1} = 0$$

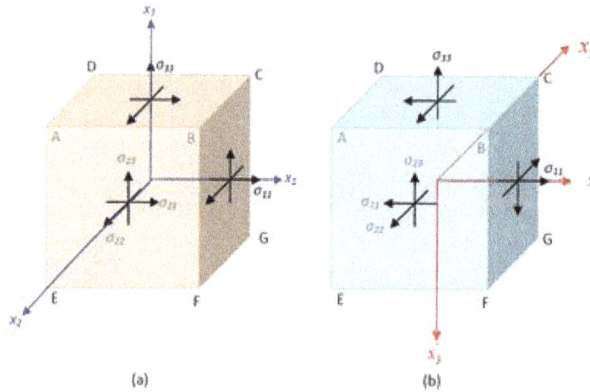

State of stress (a) in x_1, x_2, x_3 system (b) with x_1-x_2 and x_1-x_3 planes of symmetry

From this, the strains in transformed coordinate system are given as:

$$\varepsilon_{11}' = \varepsilon_{11}$$
$$\varepsilon_{22}' = \varepsilon_{22} \cos^2 \alpha + 2\varepsilon_{23} \cos \alpha \sin \alpha + \varepsilon_{33} \sin^2 \alpha$$
$$\varepsilon_{33}' = \varepsilon_{22} \sin^2 \alpha - 2\varepsilon_{23} \cos \alpha \sin \alpha + \varepsilon_{33} \cos^2 \alpha$$
$$\varepsilon_{23}' = (\varepsilon_{33} - \varepsilon_{22}) \cos \alpha \sin \alpha + \varepsilon_{23} (\cos^2 \alpha - \sin^2 \alpha)$$
$$\varepsilon_{13}' = -\varepsilon_{12} \sin \alpha + \varepsilon_{13} \cos \alpha$$
$$\varepsilon_{12}' = \varepsilon_{12} \cos \alpha + \varepsilon_{13} \sin \alpha$$

(31)

Here, it is to be noted that the shear strains are the tensorial shear strain terms.

For any angle α,

$$\varepsilon_{22} + \varepsilon_{33} = \varepsilon_{22}' + \varepsilon_{33}',$$
$$\varepsilon_{22}\varepsilon_{33} - (\varepsilon_{23})^2 = \varepsilon_{22}'\varepsilon_{33}' - (\varepsilon_{23}')^2$$
$$(\varepsilon_{12})^2 + (\varepsilon_{13})^2 = (\varepsilon_{12}')^2 + (\varepsilon_{13}')^2,$$
$$|\varepsilon_{ij}| = |\varepsilon_{ij}'|$$

(32)

and therefore, W must reduce to the form

$$W = W\left(\varepsilon_{22} + \varepsilon_{33}, \varepsilon_{22}\varepsilon_{33} - \varepsilon_{23}^2, \varepsilon_{33}, \varepsilon_{12}^2 + \varepsilon_{13}^2, |\varepsilon_{ij}|\right) \tag{33}$$

Then, for W to be invariant we must have

$$W\left(\varepsilon_{11}\varepsilon_{22} + \varepsilon_{33}, \varepsilon_{22}\varepsilon_{33} - \varepsilon_{23}^2, \varepsilon_{12}^2 + \varepsilon_{13}^2, |\varepsilon_{ij}|\right)$$
$$= W\left(\varepsilon_{11}', \varepsilon_{22}' + \varepsilon_{33}', \varepsilon_{22}'\varepsilon_{33}' - \left(\varepsilon_{23}'\right)^2, \left(\varepsilon_{12}'\right)^2 + \left(\varepsilon_{13}'\right)^2, |\varepsilon_{ij}'|\right)$$

Now, let us write the left hand side of the above equation using the C_{ij} matrix as given in Equation (26) and engineering shear strains. In the following we do some rearrangement as

$$W = \left[C_{11}\varepsilon_{11}^2\right] + \left[2\varepsilon_{11}\left(C_{12}\varepsilon_{22} + C_{13}\varepsilon_{33}\right)\right] + \left[4C_{55}\varepsilon_{13}^2 + 4C_{66}\varepsilon_{12}^2\right] +$$
$$\left[C_{22}\varepsilon_{22}^2 + C_{33}\varepsilon_{33}^2 + 2C_{23}\varepsilon_{22}\varepsilon_{33} + 4C_{44}\varepsilon_{23}^2\right]$$

Similarly, we can write the right hand side of previous equation using rotated strain components. Now, for W to be invariant it must be of the form as in Equation (33).

1. If we observe the terms containing $\left(\varepsilon_{11}\right)^2$ and $\left(\varepsilon_{11}'\right)^2$ in the first bracket, then we conclude that C_{11} is unchanged.

2. Now compare the terms in the second bracket. If we have $C_{12} = C_{13}$ then the first of Equation (32) is satisfied.

3. Now compare the third bracket. If we have $C_{55} = C_{66}$, then the third of Equation (32) is satisfied.

4. Now for the fourth bracket we do the following manipulations. Let us assume that $C_{22} = C_{33}$ and C_{23} is unchanged. Then we write the terms in fourth bracket as

$$C_{22}\left(\varepsilon_{22} + \varepsilon_{33}\right)^2 - 2C_{22}\varepsilon_{22}\varepsilon_{33} + 2C_{23}\varepsilon_{22}\varepsilon_{33} + 4C_{44}\varepsilon_{23}^2$$
$$= C_{22}\left(\varepsilon_{22} + \varepsilon_{33}\right)^2 - 2\left[\left(C_{22} - C_{23}\right)\varepsilon_{22}\varepsilon_{33} + 2C_{44}\varepsilon_{23}^2\right]$$

To have W to be invariant we need to have $C_{44} = \dfrac{C_{22} - C_{23}}{2}$ so that the third of Equation (32) is satisfied.

Thus, for transversely isotropic material (in plane x_2-x_3) the stiffness matrix becomes

$$\begin{bmatrix} C_{11} & C_{12} & C_{12} & 0 & 0 & 0 \\ & C_{22} & C_{23} & 0 & 0 & 0 \\ & & {}_{22} & 0 & 0 & 0 \\ & & & -\left(C_{22} \quad C_{23}\right) & 0 & 0 \\ & Symmetric & & & C_{66} & \\ & & & & & {}_{66} \end{bmatrix} \tag{34}$$

Thus, there are only 5 independent elastic constants for a transversely isotropic material.

Second Approach: Comparison of Constants

This can also be verified from the elastic constants expressed in terms of engineering constants like E, v and G. Recall the constitutive equation for orthotropic material expressed in terms of engineering constants. For the transversely isotropic materials the following relations hold.

$$E_2 = E_3, \quad v_{12} = v_{13}$$

$$G_{12} = G_{13}, \quad G_{23} = \frac{E_2}{2(1 + v_{23})}$$

When these relations are used in the constitutive equation for orthotropic material expressed in terms of engineering constants, the stiffness matrix relations in Equation (34) are verified.

Isotropic Bodies

If the function W remains unaltered in form under all possible changes to other rectangular Cartesian systems of axes, the body is said to be Isotropic. In this case, W is a function of the strain invariants. W must be unaltered in form under the transformations

$$x_1' = x_1 \cos \alpha + x_2 \sin \alpha$$
$$x_2' = -x_1 \sin \alpha + x_2 \cos \alpha$$
$$x_3' = x_3 \tag{35}$$

and

$$x_3' = x_3 \cos \alpha + x_1 \sin \alpha$$
$$x_1' = -x_3 \sin \alpha + x_1 \cos \alpha$$
$$x_2' = x_2 \tag{36}$$

In other words, W when expressed in terms of ε_{ij}' must be obtained from Equation (33) simply by replacing ε_{ij} by ε_{ij}'. It is seen that for this to be true under the transformation Equation (35). We can write

$$\frac{\partial x_1'}{\partial x_1} = \frac{\partial x_2'}{\partial x_2} = \cos \alpha, \frac{\partial x_1'}{\partial x_2} = -\frac{\partial x_2'}{\partial x_1} = \sin \alpha,$$

$$\frac{\partial x_1'}{\partial x_3} = \frac{\partial x_2'}{\partial x_3} = \frac{\partial x_3'}{\partial x_1} = \frac{\partial x_3'}{\partial x_2} = 0, \frac{\partial x_3'}{\partial x_3} = 1$$

And the transformed strains are given as

$$\varepsilon'_{11} = \varepsilon_{11}\cos^2\alpha + 2\varepsilon_{12}\sin\alpha\cos\alpha + \varepsilon_{33}\sin^2$$

$$\varepsilon'_{22} = \varepsilon_{11}\sin^2\alpha - 2\varepsilon_{12}\sin\alpha\cos\alpha + \varepsilon_{22}\cos^2\alpha$$

$$\varepsilon'_{33} = \varepsilon_{33}$$

$$\varepsilon'_{23} = -\varepsilon_{13}\sin\alpha + \varepsilon_{23}\cos\alpha$$

$$\varepsilon'_{13} = \varepsilon_{13}\cos\alpha + \varepsilon_{23}\sin\alpha$$

$$\varepsilon'_{12} = (\varepsilon_{22} - \varepsilon_{11})\sin\alpha\cos\alpha + \varepsilon_{12}(\cos^2\alpha - \sin^2\alpha) \tag{37}$$

Thus, for any angle α,

$$\varepsilon_{11} + \varepsilon_{22} = (\varepsilon'_{11}) + \varepsilon'_{22},$$

$$\varepsilon_{11}\varepsilon_{22} - (\varepsilon_{12})^2 + \varepsilon'_{11}\varepsilon'_{22} - (\varepsilon'_{12})^2,$$

$$(\varepsilon_{13})^2 + (\varepsilon_{23})^2 = (\varepsilon'_{13})^2 + (\varepsilon'_{23})^2,$$

$$|\varepsilon_{ij}| = |\varepsilon'_{ij}| \tag{38}$$

and therefore, W must reduce to the form

$$W = W\left(\varepsilon_{11} + \varepsilon_{22}, +\varepsilon_{11}\varepsilon_{22} - \varepsilon_{12}^2, \varepsilon_{33}, \varepsilon_{13}^2 + \varepsilon_{23}^2, |\varepsilon_{ij}|\right) \tag{39}$$

Then, for W to be invariant we must have

$$W\left(\varepsilon_{11} + \varepsilon_{22}, +\varepsilon_{11}\varepsilon_{22} - \varepsilon_{12}^2, \varepsilon_{33}, \varepsilon_{13}^2 + \varepsilon_{23}^2, |\varepsilon_{ij}|\right)$$

$$= W\left(\varepsilon'_{11} + \varepsilon'_{22}, +\varepsilon'_{11}\varepsilon'_{22} - (\varepsilon'_{12})^2, \varepsilon'_{33}(\varepsilon'_{13})^2 + (\varepsilon'_{23})^2, |\varepsilon'ij|\right)$$

Now, let us write the left hand side of above equation using the C_{ij} matrix as given in Equation (34) and engineering shear strains. In the following we do some rearrangement as:

$$W = [C_{22}\varepsilon_{33}^2] + [2\varepsilon_{33}(C_{12}\varepsilon_{11} + C_{23}\varepsilon_{22})] + \left[\frac{4}{2}(C_{22} - C_{23})\varepsilon_{23}^2 + 4C_{66}\varepsilon_{13}^2\right] +$$

$$[C_{11}\varepsilon_{11}^2 + C_{22}\varepsilon_{22}^2 + 2C_{12}\varepsilon_{11}\varepsilon_{12} + 4C_{66}\varepsilon_{12}^2] \tag{40}$$

Similarly, we can write the right hand side of the previous equation using rotated strain components. Now, for W to be invariant it must be of the form as in Equation (39)

1. From the second bracket, if we propose $C_{23} = C_{12}$, then we can satisfy the first of Equation (38).

2. From the third bracket, third of Equation (38) holds true when $C_{66} = \dfrac{C_{22} - C_{23}}{2} = \dfrac{C_{22} - C_{12}}{2} = C_{44}$.

3. The fourth bracket is manipulated as follows:

$$C_{11}\varepsilon_{11}^2 + C_{22}\varepsilon_{22}^2 + 2C_{12}\varepsilon_{11}\varepsilon_{22} + 4C_{66}\varepsilon_{12}^2$$

$$= C_{11}\left(\varepsilon_{11}\varepsilon_{22}\right)^2 - 2C_{11}\varepsilon_{11}\varepsilon_{22} + 2C_{12}\varepsilon_{11}\varepsilon_{22} + 4C_{66}\varepsilon_{12}^2$$

$$= C_{11}\left(\varepsilon_{11} + \varepsilon_{22}\right)^2 - 2\left[\left(C_{11} - C_{12}\right)\varepsilon_{11}\varepsilon_{22} - 2C_{66}\varepsilon_{12}^2\right]$$

Thus, to satisfy the second of Equation (38) we must have $C_{66} = \dfrac{C_{11} - C_{12}}{2}$. Further, we should have $C_{22} = C_{11}$. From our observation in 2, we can write $C_{44} = C_{55} = C_{66} = \dfrac{C_{11} - C_{12}}{2}$.

It follows automatically that W is unaltered in form under the transformation in Equation (36).

Thus, the stiffness matrix for isotropic material becomes as

$$C_{ij} = \begin{bmatrix} C_{11} & C_{12} & C_{12} & 0 & 0 & 0 \\ & C_{11} & C_{12} & 0 & 0 & 0 \\ & & C_{11} & 0 & 0 & 0 \\ & & & \frac{1}{2}(C_{11} - C_{12}) & 0 & 0 \\ & Symmetric & & & \frac{1}{2}(C_{11} - C_{12}) & 0 \\ & & & & & \frac{1}{2}(C_{11} - C_{12}) \end{bmatrix} \tag{41}$$

Thus, for an isotropic material there are only two independent elastic constants. It can be verified that W is unaltered in form under all possible changes to other rectangular coordinate systems, that is, it is the same function of ε_{ij}' as it is of ε_{ij} when x_i is changed to x_i'.

Stiffness Transformation

It is required to relate the stress components with strain components in global *xyz* directions. The stiffness matrix which relates the stress and strain components in global directions is called as transformed stiffness matrix. We will derive an expression for the transformed stiffness matrix as follows.

The constitutive equation in principal material coordinates, as given in Equation (11), is

$$\{\sigma\}_{123} = [C]\{\varepsilon\}_{123} \tag{42}$$

Now, we express $\{\sigma\}_{123}$ *and* $\{\varepsilon\}_{123}$ using equation to transform stresses and to transform strains. Substituting these equations, we get

$$[T]\{\sigma\}_{xyz} = [C][T_2]\{\varepsilon\}_{xyz} \tag{43}$$

Pre-multiplying both sides by $[T_1]^{-1}$, we get

$$\{\sigma\}_{xyz} = [T_1]^{-1}[C][T_2]\{\varepsilon\}_{xyz}$$
$$\{\sigma\}_{xyz} = [\bar{C}]\{\varepsilon\}_{xyz} \tag{44}$$

where we define the transformed stiffness matrix $[\bar{C}]$ as

$$[\bar{C}] \quad [T_1] \quad [C][T_2] \tag{45}$$

The transformation matrices $[T_1]$ and $[T_2]$ can be inverted as follows

$$[T_i(\theta)]^{-1} = [T_i(-\theta)] \quad i = 1,2 \tag{46}$$

The final form of the transformed stiffness matrix is given in Equation (47).

$$[\bar{C}] = \begin{bmatrix} \bar{C}_{11} & \bar{C}_{12} & \bar{C}_{13} & 0 & 0 & \bar{C}_{16} \\ \bar{C}_{12} & \bar{C}_{22} & \bar{C}_{23} & 0 & 0 & \bar{C}_{26} \\ \bar{C}_{13} & \bar{C}_{23} & \bar{C}_{33} & 0 & 0 & \bar{C}_{36} \\ 0 & 0 & 0 & \bar{C}_{44} & \bar{C}_{45} & 0 \\ 0 & 0 & 0 & \bar{C}_{45} & \bar{C}_{55} & 0 \\ \bar{C}_{16} & \bar{C}_{26} & \bar{C}_{36} & 0 & 0 & \bar{C}_{66} \end{bmatrix} \tag{47}$$

The individual $\bar{C}_{ij}$ terms of this matrix are determined using $[T_1]^{-1}, [T_2]$ and relation for $[\bar{C}]$. The individual terms are given in Equation (48).

Note: The transformed stiffness matrix is symmetric in nature.

Note: The transformed stiffness matrix given in Equation (47) has exactly the same form as a stiffness matrix for a monoclinic material. Thus, we can conclude that a transformation through an arbitrary angle θ about direction 3, leads to a monoclinic material behaviour.

The same can be seen from the plane of elastic symmetry considerations in *xyz* coordinate system. The given lamina is symmetric only about *xy* plane. Thus, the transformed stiffness matrix in Equation (47) is consistent with monoclinic material.

Note: Transformed stiffness coefficient terms are fourth order in the sine and cosine functions. It is very important to use appropriate precision level while calculating (in examinations and writing computer codes) these coefficients.

$$\bar{C}_{11} = m^4 C_{11} + 2m^2 n^2 (C_{12} + 2C_{66}) + n^4 C_{22}$$
$$\bar{C}_{12} = n^2 m^2 (C_{11} + C_{22} - 4C_{66}) + (n^4 + m^4)C_{12}$$
$$\bar{C}_{13} = m^2 C_{13} + n^2 C_{23}$$
$$\bar{C}_{16} = nm[m^2 (C_{11} - C_{12} - 2C_{66}) + n^2 (C_{12} - C_{22} + 2C_{66})]$$

$$\bar{C}_{22} = n^4 C_{11} + 2m^2 n^2 \left(C_{12} + 2C_{66}\right) + m^4 C_{22}$$

$$\bar{C}_{23} = n^2 C_{13} + m^2 C_{23}$$

$$\bar{C}_{26} = nm\left[n^2 \left(C_{11} - C_{12} - 2C_{66}\right) + m^2 \left(C_{12} - C_{22} + 2C_{66}\right)\right]$$

$$\bar{C}_{33} = C_{33}$$

$$\bar{C}_{36} = mn\left(C_{13} - C_{23}\right)$$

$$\bar{C}_{44} = m^2 C_{44} + n^2 C_{55}$$

$$\bar{C}_{45} = mn\left(C_{55} - C_{44}\right)$$

$$\bar{C}_{55} = n^2 C_{44} + m^2 C_{55}$$

$$\bar{C}_{66} = n^2 m^2 \left(C_{11} - 2C_{12} + C_{22}\right) + \left(n^2 - m^2\right)^2 C_{66} \tag{48}$$

The constitutive equation $\{\sigma\}_{xyz} = \left[\bar{C}\right]\{\varepsilon\}_{xyz}$ becomes

$$
\begin{Bmatrix} \sigma_{xx} \\ \sigma_{yy} \\ \sigma_{xx} \\ \tau_{yz} \\ \tau_{xz} \\ \tau_{xy} \end{Bmatrix} =
\begin{bmatrix}
\bar{C}_{11} & \bar{C}_{12} & \bar{C}_{13} & 0 & \bar{C}_{16} \\
\bar{C}_{12} & \bar{C}_{22} & \bar{C}_{23} & 0 & \bar{C}_{26} \\
\bar{C}_{13} & \bar{C}_{23} & \bar{C}_{33} & 0 & \bar{C}_{36} \\
0 & 0 & 0 & \bar{C}_{44} & \bar{C}_{45} \\
0 & 0 & 0 & \bar{C}_{45} & \bar{C}_{55} \\
\bar{C}_{16} & \bar{C}_{26} & \bar{C}_{36} & 0 & \bar{C}_{66}
\end{bmatrix}
\begin{Bmatrix} \varepsilon_{xx} \\ \varepsilon_{yy} \\ \varepsilon_{xx} \\ \gamma_{yz} \\ \gamma_{xz} \\ \gamma_{xy} \end{Bmatrix}
$$

Compliance Transformation

We are going to follow a procedure to transform compliance matrix similar to one used for transformation of a stiffness matrix. We have the constitutive equation in principal material direction as in Equation (42). We can write this in inverted form as

$$\{\varepsilon\}_{123} = [S]\{\sigma\}_{123} \tag{49}$$

using Equation we get

$$[T_2]\{\varepsilon\}_{xyz} = [S][T_1]\{\sigma\}_{xyz} \tag{50}$$

Pre-multiplying both sides by $[T_2]^{-1}$, we get

$$\{\varepsilon\}_{xyz} = [T_2]^{-1}[S][T_1]\{\sigma\}_{xyz}$$

$$\{\varepsilon\}_{xyz} = [\bar{S}]\{\sigma\}_{xyz} \tag{51}$$

where, we define the transformed compliance matrix $[\bar{S}]$ as

$$[\bar{S}] = [T_2]^{-1}[S][T_1] \tag{52}$$

Alternately, we can find $\left[\bar{S}\right]$ by inverting the transformed stiffness matrix $\left[\bar{C}\right]$. Thus, inverting $\left[\bar{C}\right]$ from Equation (45), we get

$$\left[\bar{S}\right] = \left[\bar{C}\right]^{-1} = \left(\left[T_1\right]^{-1}\left[C\right]\left[T_2\right]\right)^{-1}$$

$$= \left[T_2\right]^{-1}\left[C\right]^{-1}\left(\left[T_1\right]^{-1}\right)^{-1}$$

$$= \left[T_2\right]^{-1}\left[S\right]\left[T_1\right]$$

After carrying out the calculation for $\left[\bar{S}\right]$, it is easy to give its form as follows

$$\left[\bar{S}\right] = \begin{bmatrix} \bar{S}_{11} & \bar{S}_{12} & \bar{S}_{13} & 0 & 0 & \bar{S}_{16} \\ \bar{S}_{12} & \bar{S}_{22} & \bar{S}_{23} & 0 & 0 & \bar{S}_{26} \\ \bar{S}_{13} & \bar{S}_{23} & \bar{S}_{33} & 0 & 0 & \bar{S}_{36} \\ 0 & 0 & 0 & \bar{S}_{44} & \bar{S}_{45} & 0 \\ 0 & 0 & 0 & \bar{S}_{45} & \bar{S}_{55} & 0 \\ \bar{S}_{16} & \bar{S}_{26} & \bar{S}_{36} & 0 & 0 & \bar{S}_{66} \end{bmatrix} \qquad (53)$$

Note that $\left[\bar{S}\right]$ has the same symmetric form as the transformed stiffness matrix.

The individual terms of the compliance matrix are obtained by carrying out multiplication of matrices as in Equation (53) and are given below.

$$\bar{S}_{11} = m^4 S_{11} + m^2 n^2 \left(2S_{12} + S_{66}\right) + n^4 S_{22}$$

$$\bar{S}_{12} = n^2 m^2 \left(S_{11} + S_{22} - S_{66}\right) + \left(n^4 + m^4\right)S_{12}$$

$$\bar{S}_{13} = m^2 S_{13} + n^2 S_{23}$$

$$\bar{S}_{16} = nm\left[m^2\left(2S_{11} - 2S_{12} - S_{66}\right) + n^2\left(2S_{12} - 2S_{22} + S_{66}\right)\right]$$

$$\bar{S}_{22} = n^4 S_{11} + m^2 n^2 \left(2S_{12} + S_{66}\right) + m^4 S_{22}$$

$$\bar{S}_{23} = n^2 S_{13} + m^2 S_{23}$$

$$\bar{S}_{26} = nm\left[n^2\left(2S_{11} - 2S_{12} - S_{66}\right) + m^2\left(2S_{12} - 2S_{22} + S_{66}\right)\right]$$

$$\bar{S}_{33} = S_{33}$$

$$\bar{S}_{36} = 2mn\left(S_{13} - S_{23}\right)$$

$$\bar{S}_{44} = m^2 S_{44} + n^2 S_{55}$$

$$\bar{S}_{45} = mn\left(S_{55} - S_{44}\right)$$

$$\bar{S}_{55} = n^2 S_{44} + m^2 S_{55}$$

$$\bar{S}_{66} = 4n^2 m^2 \left(S_{11} - 2S_{12} + S_{22}\right) + \left(n^2 - m^2\right)S_{66}$$

Thermal Effects

Thermal effects (effects due to change in temperature) are very important in composite materials for various reasons. The analysis of composites with thermal effects and effective thermal properties of the composite are two of the main reasons.

Important Issues from Analysis Point of View

1. The composite materials are used in environment where thermal gradients are unavoidable. For example, the helicopter containing composite fuselage operates at -50° C during winter at Leh and same helicopter can operate at +50° C during summer in the desert of Rajasthan. Thus, the effect of temperature gradient on the service performance of the composite is very important. In such service conditions, the layers of composite material tend to expand or contract but are restricted due to adjacent layers. Thus, it induces thermal stresses.

2. Most of the fabrication processes of polymer matrix composites have thermal cycles for matrix curing. A typical cycle involves raising the temperature to a certain level and holding it there for specified time and bringing it back to room temperature. It is well known that the fibre and matrix materials have different coefficients of thermal expansion (defined below). This mismatch produces residual thermal stresses because the fibres and matrix material are constrained in a composite.

Important Issues from Effective Thermal Properties Point of View

The second reason for the study of thermal effects is the effective properties of the composite materials.

1. Finding effective thermal properties of the composite theoretically to get an estimate requires sophisticated mathematical modeling when one considers:

    a. Difference in coefficients of thermal expansion of fibre and matrix materials

    b. The direction dependence of coefficients of thermal expansion in these materials

    c. Curing cycle temperature variations. This point is important because for some of the materials the coefficient of thermal expansion changes with temperature.

2. Finding the effective thermal properties for lamina in global direction with oriented fibres as shown in Figure requires a special attention.

Further, finding these effective properties by laboratory test is also a challenge. Thus, for the various reasons mentioned above the study of thermal effect is very important. In the following, we develop a systematic way to handle effective thermal properties of a lamina along global directions.

It is well known that when a material is subjected to thermal gradient, it undergoes a deformation. The strain due to thermal changes is called thermal strain (denoted by superscript (T)). In general, the thermal strain is proportional to the temperature change ΔT. The constant of proportionality is called coefficient of thermal expansion. Thus, we can write the thermal strains in principal material directions for an orthotropic material as

$$\left\{\varepsilon^{(T)}\right\}_{123} = \left\{\alpha\right\}_{123} \Delta T \tag{54}$$

where $\left\{\alpha\right\}_{123} = \left\{\alpha_1, \alpha_2, \alpha_3 0, 0, 0\right\}^T$ denote the coefficients of thermal expansion in principal material directions. It should be noted that for an orthotropic material in principal directions there are no shear strains due to thermal effects like in an isotropic material. For an isotropic material the coefficient of thermal expansion is same in any direction. However, for an orthotropic material $\alpha_1 \gg \alpha_2, \alpha_3$. The thermal expansion of an elemental cube in principal directions for an isotropic and orthotropic material is shown in Figure.

These thermal strains will not produce stresses unless these are constrained. The thermal strains which do not produce stresses are known as free thermal strains. However, in case of composites the fibres and matrix are constrained in a lamina and layers are constrained in a laminate. Thus, in composite the thermal strains produce the thermal stresses.

The thermal strains are given in principal material directions as given in Equation (54). Let us consider that we need to find these strains in a global coordinate system. We need to transform them from 123 coordinate system to xyz coordinate system by a rotation θ about 3-axis. Thus, similar to Equation, we can write:

$$\left\{\varepsilon^{(T)}\right\}_{123} = \left[T_2\right]\left\{\varepsilon^{(T)}\right\}_{xyz}$$

$$\left\{\varepsilon^{(T)}\right\}_{xyz} = \left[T_2\right]^{-1}\left\{\varepsilon^{(T)}\right\}_{123} \tag{55}$$

Substituting Equation (54) in the above equation,

$$\left\{\varepsilon^{(T)}\right\}_{xyz} = \left[T_2\right]^{-1}\left\{\alpha\right\}_{123} \Delta T$$

$$= \left\{\alpha\right\}_{xyz} \Delta T \tag{56}$$

where

$$\left\{\alpha\right\}_{xyz} = \left[T_2\right]^{-1}\left\{\alpha\right\}_{123} \quad \left\{\alpha\right\}_{xyz} = \left\{\alpha_{xx} \alpha_{yy} \alpha_{zz} 0 0 \alpha_{xy}\right\}^T \text{ and } \left\{\alpha\right\}_{123} = \left\{\alpha_1 \alpha_2 \alpha_3 0 0 0\right\}^T \tag{57}$$

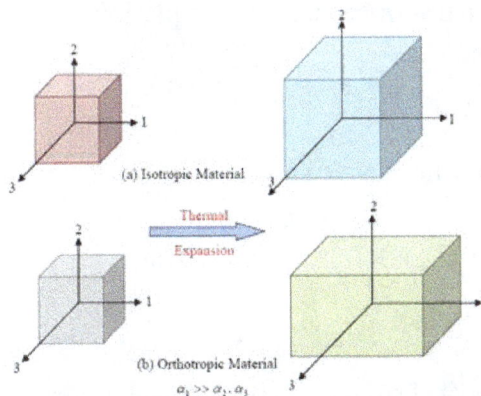

(a) Isotropic Material

Thermal
Expansion

(b) Orthotropic Material
$\alpha_1 \gg \alpha_2, \alpha_3$

Thermal expansion in an isotropic and orthotropic material

On substitution of $\left[T_2\right]^{-1}$ in the above equation, we get the following individual terms of coefficients thermal expansion in xyz directions.

$$\alpha_{xx} = m^2\,\alpha_1 + n^2\,\alpha_2$$
$$\alpha_{yy} = n^2\,\alpha_1 + m^2\,\alpha_2$$
$$\alpha_{zz} = \alpha_3$$
$$\alpha_{xy} = 2mn\left(\alpha_1 - \alpha_2\right)$$

$$(58)$$

Using Equation (58) in Equation (56), the engineering thermal strains in global coordinates are given as

$$\begin{Bmatrix} \varepsilon_{xx}^{(T)} \\ \varepsilon_{yy}^{(T)} \\ \varepsilon_{zz}^{(T)} \\ \gamma_{yz}^{(T)} \\ \gamma_{xz}^{(T)} \\ \gamma_{xy}^{(T)} \end{Bmatrix} = \begin{Bmatrix} m^2\alpha_1 + n^2\alpha_2 \\ n^2\alpha_1 + m^2\alpha_2 \\ \alpha_3 \\ 0 \\ 0 \\ 2mn\left(\alpha_1 - \alpha_2\right) \end{Bmatrix} \Delta T$$

$$(59)$$

Thus, from this equation it should be noted that the transformation of thermal strains in global coordinates gives normal strain components and a shear strain component in xy plane for an orthotropic material with $\alpha_1 \neq \alpha_2$ and fiber orientations other than $0°\ and\ 90°$.

Thermo-Elastic Constitutive Equation

Let us assume that the total strain, $\{\varepsilon\}$ in a composite is a superposition of the free thermal strain $\{\varepsilon^{(T)}\}$ and the strain due to mechanical loads (also known as mechanical strains) $\{\varepsilon^{(\sigma)}\}$. Thus,

$$\{\varepsilon\}_{123} = \{\varepsilon^{(\sigma)}\}_{123} + \{\varepsilon^{(T)}\}_{123}$$

Now, for mechanical strains we use the constitutive equation as

$$\{\varepsilon^{(\sigma)}\}_{123} + [S]\{\sigma\}_{123}$$

Thus, we can write the total thermo-elastic strain as

$$\{\varepsilon\}_{123} = [S]\{\sigma\}_{123} + \{\varepsilon^{(T)}\}_{123}$$

$$(60)$$

Equation (60) can be written as

$$[S]\{\sigma\}_{123} = \{\varepsilon\}_{123} - \{\varepsilon^{(T)}\}_{123}$$

Premultiplying the above expression by $[S]^{-1}$, we get the stresses as

$$\{\sigma\}_{123} = [S]^{-1}\left(\{\varepsilon\}_{123} - \{\varepsilon^{(T)}\}_{123}\right)$$

$$\{\sigma\}_{123} = [C]\left(\{\varepsilon\}_{123} - \{\varepsilon^{(T)}\}_{123}\right) \tag{61}$$

Equation (61) is the basic constitutive equation for thermo-elastic stress analysis.

Using Equation (61) and similar to Equation (44) we can find the stresses due to thermo-elastic effects in global directions as,

$$\{\sigma\}_{xyz} = \{\bar{C}\}\left(\{\varepsilon\}_{xyz} - \{\varepsilon^{(T)}\}_{xyz}\right) \tag{62}$$

where

$$\{\varepsilon\}_{xyz} - \{\varepsilon^{(T)}\}_{xyz} = \left\{\varepsilon_{xx} - \varepsilon_{xx}^{(T)}\varepsilon_{yy} - \varepsilon_{yy}^{(T)}\varepsilon_{xx} - \varepsilon_{xx}^{(T)}\gamma_{yz}\,\gamma_{xz}\,\gamma_{xy} - \gamma_{xy}^{(T)}\right\}^T$$

Equation (62) is inverted to give the total strains in terms of the mechanical and free thermal strains as

$$\{\varepsilon\}_{xyz} = \lfloor\bar{S}\rfloor\{\sigma\}_{xyz} + \{\varepsilon^{(T)}\}_{xyz} \tag{63}$$

Effect of Moisture

The polymer matrix composite materials, during their service can absorb moisture from the environment. The effect of absorption of moisture is to degrade the various material properties of the composite. Further, this results in an expansion. It is called hygroscopic expansion. However, this expansion is again constrained as in thermal expansion. Hence, when dealing with the hygroscopic expansions, a treatment similar to thermal expansion is used.

The hygroscopic strains $\{\varepsilon^{(H)}\}$ are assumed to be proportional to the percentage moisture absorbed, ΔM This percentage is measured in terms of weight of the moisture. The constant of proportionality, $\{\beta\}$ is the coefficient of hygroscopic expansion.

Thus, in principal coordinates the hygroscopic strains are

$$\{\varepsilon^{(H)}\}_{123} = \{\beta\}_{123}\,\Delta M \tag{64}$$

where

$$\{\beta\}_{123} = \{\beta_1\,\beta_2\,\beta_3\,0\,0\,0\}^T \tag{65}$$

denotes the coefficients of hygroscopic expansion in principal material directions. Following a similar procedure for thermal strains, we can write strains due to hygroscopic expansion in *xyz* coordinates as

$$\left\{\varepsilon^{(H)}\right\}_{xyz} = \left\{\beta\right\}_{xyz} \Delta M$$

(66)

We can write $\left\{\beta\right\}_{xyz} = \left\{\beta_{xx} \ \beta_{yy} \ \beta_{zz} \ 0 \ 0 \ \beta_{xy}\right\}^{T}$ using values of in 123 directions as $\left\{\beta\right\}_{xyz} = \left[T_2\right]^{-1}\left\{\beta\right\}_{123}$.

Thus, comparing Equation (56) and Equation (66), it is easy to conclude that the coefficients of hygroscopic expansion will vary similar to the coefficients of thermal expansion as a function of orientation of fibres.

Hygro-Thermo-Elastic Constitutive Equation

This is the most general formulation for the mechanical, thermal and hygral effects on stress analysis in composites. Here, we superimpose the strains due to these three effects to give us the total strain as

$$\left\{\varepsilon\right\}_{123} = \left\{\varepsilon^{(\sigma)}\right\}_{123} + \left\{\varepsilon^{(T)}\right\}_{123} + \left\{\varepsilon^{(H)}\right\}_{123}$$

(67)

Using constitutive equation for mechanical strains, we get

$$\left\{\varepsilon\right\}_{123} = \left[S\right]\left\{\sigma\right\}_{123} + \left\{\varepsilon^{(T)}\right\}_{123} + \left\{\varepsilon^{(H)}\right\}_{123}$$

(68)

The stresses in the composite can be given as

$$\left\{\sigma\right\}_{123} = \left[C\right]\left(\left\{\varepsilon\right\}_{123} - \left\{\varepsilon^{(T)}\right\}_{123} - \left\{\varepsilon^{(H)}\right\}_{123}\right)$$

(69)

These stresses in global coordinates xyz can be written as

$$\left\{\sigma\right\}_{xyz} = \left[\bar{C}\right]\left(\left\{\varepsilon\right\}_{xyz} - \left\{\varepsilon^{(T)}\right\}_{xyz} - \left\{\varepsilon^{(H)}\right\}_{xyz}\right)$$

(70)

where

$$\left\{\varepsilon\right\}_{xyz} - \left\{\varepsilon^{(T)}\right\}_{xyz} - \left\{\varepsilon^{(H)}\right\}_{xyz} = \begin{Bmatrix} \varepsilon_{xx} - \varepsilon_{xx}^{(T)} - \varepsilon_{xx}^{(H)} \\ \varepsilon_{yy} - \varepsilon_{yy}^{(T)} - \varepsilon_{yy}^{(H)} \\ \varepsilon_{zz} - \varepsilon_{zz}^{(T)} - \varepsilon_{zz}^{(H)} \\ yz \\ xz \\ \gamma_{xy} - \gamma_{xy}^{(T)} - \gamma_{xy}^{(H)} \end{Bmatrix}$$

(71)

Examples

Example 2: Transform the stiffness and compliance matrix of Example 1 about axis 3 by an angle of $\theta = 30°$.

Solution

Approach 1: One can find the transformation matrices $[T_1]$ *and* $[T_2]$ and do the matrix multiplication as given in Equation (45) for transformed stiffness matrix and then inverse this matrix or do the matrix multiplication as given in Equation (52) to get the transformed compliance matrix. The use of Equation (45) and Equation (52) is suggested because remembering $[T_1]$ *and* $[T_2]$ is not so difficult. Further, their inverse can be easily found with the help of Equation (45).

For

$$\theta = 30°, m = \cos 30° = 03.86603, n = \sin 30° = 0.5$$

Thus

$$[T_1] = \begin{bmatrix} 0.75 & 0.25 & 0 & 0 & 0 & 0.86603 \\ 0.25 & 0.75 & 0 & 0 & 0 & -0.86603 \\ 0 & 0 & 1 & 0 & 0 & 0 \\ 0 & 0 & 0 & 0.86603 & -0.5 & 0 \\ 0 & 0 & 0 & 0.5 & 0.86603 & 0 \\ -0.43301 & 0.43301 & 0 & 0 & 0 & 0.5 \end{bmatrix}$$

$$[T_2] = \begin{bmatrix} 0.75 & 0.25 & 0 & 0 & 0 & 0.43301 \\ 0.25 & 0.75 & 0 & 0 & 0 & -0.43301 \\ 0 & 0 & 1 & 0 & 0 & 0 \\ 0 & 0 & 0 & 0.86603 & -0.5 & 0 \\ 0 & 0 & 0 & 0.5 & 0.86603 & 0 \\ -086603 & 0.86603 & 0 & 0 & 0 & 0.5 \end{bmatrix}$$

$$[T_1]^{-1} = \begin{bmatrix} 0.75 & 0.25 & 0 & 0 & 0 & -0.86603 \\ 0.25 & 0.75 & 0 & 0 & 0 & 0.86603 \\ 0 & 0 & 1 & 0 & 0 & 0 \\ 0 & 0 & 0 & 0.86603 & 0.5 & 0 \\ 0 & 0 & 0 & -0.5 & 0.86603 & 0 \\ 0.43301 & -0.43301 & 0 & 0 & 0 & 0.5 \end{bmatrix}$$

$$[T_2]^{-1} = \begin{bmatrix} 0.75 & 0.25 & 0 & 0 & 0 & -0.43301 \\ 0.25 & 0.75 & 0 & 0 & 0 & 0.43301 \\ 0 & 0 & 1 & 0 & 0 & 0 \\ 0 & 0 & 0 & 0.86603 & 0.5 & 0 \\ 0 & 0 & 0 & -0.5 & 0.86603 & 0 \\ 086603 & -0.86603 & 0 & 0 & 0 & 0.5 \end{bmatrix}$$

$$[\bar{C}][T_2]^{-1}[C][T_2] = \begin{bmatrix} 80.282 & 25.005 & 5.303 & 0 & 0 & 36.439 \\ & 22.465 & 5.402 & 0 & 0 & 13.632 \\ & & 13.309 & 0 & 0 & -0.086 \\ & & & 4.539 & 1.157 & 0 \\ & Symmetric & & & 5.932 & 0 \\ & & & & & 26.352 \end{bmatrix}$$

Unit of all Transformed Stiffness Coefficients is *GPa*.

$$[\bar{S}] = [T_2]^{-1}[S][T_1] = \begin{bmatrix} 0.03772 & -0.01126 & -0.01075 & 0 & 0 & -0.04636 \\ & 0.07921 & -0.02782 & 0 & 0 & -0.02548 \\ & & 0.09091 & 0 & 0 & 0.02956 \\ & & & 0.22878 & -0.04462 & 0 \\ & symmetric & & & 0.17727 & 0 \\ & & & & & 0.11534 \end{bmatrix}$$

Unit of all Transformed Compliance Coefficients is *1/GPa*.

Approach 2: You can write the expanded form for transformed stiffness and compliance coefficients in Equation (3.76) and Equation (3.82). However, the readers are suggested to use this approach only when they are confident of remembering these terms.

Example 3: The coefficients of moisture absorption for T300/5208 composite material are $\beta_1 = 0.0 / wt\%, \beta_2 = \beta_3 = 6.67 \times 10^{-3} / wt\%$. Plot the variation these coefficients between $-90° < \theta < 90°$.

Solution

We have the expression for variation of the coefficients of moisture absorption as

$$\beta_{xx} = m^2 \beta_1 + n^2 \beta_2$$
$$\beta_{yy} = n^2 \beta_1 + m^2 \beta_2$$
$$\beta_{zz} = \beta_3$$
$$\beta_{xy} = 2mn(\beta_1 - \beta_2)$$

where, $m = \cos\theta$ *and* $n = \sin\theta$. We plot the above variation using a computer code. The final plot is shown in Figure.

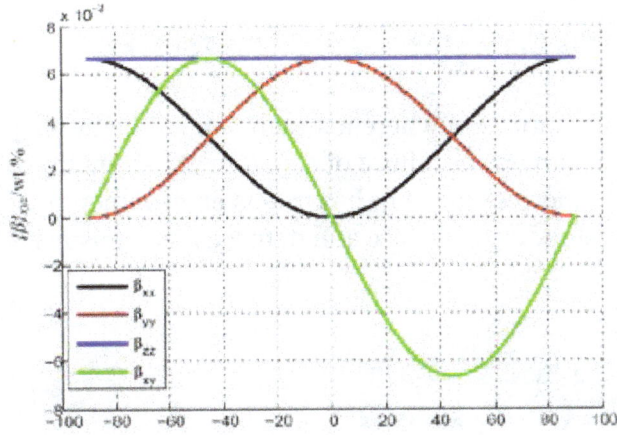

Variation of coefficients of moisture expansion with orientation of fibres

Failure and Damage Mechanisms

Failure mechanics studies the situation where a system stops functioning properly. Some of the standards used to measure a damage or failure of a system are strength, stiffness, yielding, bending, resistance to lightening and resistance to hazardous environmental agents. This section will provide an integrated understanding of failure and damage mechanisms.

Failure and Damage

Failure of a structure or a system, in general, refers to the condition that the structure or system stops functioning satisfactorily. The criteria to decide the satisfactory functioning can be subjective or quantitative. In general, in engineering applications the failure is quantified using various criteria. The following are some of the criteria used to quantify a failure:

1. Strength

2. Form failure

3. Stiffness

4. Yielding

5. Fatigue life

6. Bending

7. Corrosion resistance

8. Impact resistance

9. Resistance to lightening

10. Resistance to hazardous environmental agents

However, the list is in-exhaustive with many such criteria. In general, failure is understood as complete de-functioning of the structure.

In case of composites, the failure of a lamina or laminate needs special attention. In case of laminates there are a number of local failures before it completely breaks into two or more pieces. The local level failure is called as "damage". In case of fibrous composites the term "local" refers to the individual constituent phases – fibre and matrix. Thus, damage in case of fibrous composites is a micro level event.

It is important to note that the ultimate failure (rupture/breaking) of the laminate takes place by gradual accumulation of damage. In turn, this is manifested at the lamina or laminate level by some form of failure. Thus, the "first failure" in laminates does not mean the "final failure". The development of additional local failures with increasing loads or time is termed as "damage accumulation". The terms "damage growth" and "damage propagation" are equivalently used for damage accumulation. The branch of mechanics which deals with the study of initiation and accumulation of damage until and including complete rupture is called as "damage mechanics".

In this section we are going to see the fibre-matrix level failure mechanisms in detail. The failure at lamina/laminate or macro-level is the ultimate result of the local failures. Thus, the understanding of these mechanisms is a key point in the development of a reliable and accurate failure theory for laminated composites. Further, this understanding also helps in developing new materials with higher strength.

Defects in Composites

The following are the types of defects that generally occur in a composite:

1) Fibre-matrix debonding	8) Matrix cracking and crazing
2) Fibre misalignment	9) Density variation (due to resin distribution)
3) Cut or broken fibres	10) Improper curing of resin
4) Delamination	11) Impact damage (tool drop)
5) Inclusions	12) Abrasion and scratches
6) Voids and blisters	13) Machining problems
7) Wrinkles	

Sources of Defects and Damages in Composite:

There are two main sources which can introduce defects and/or damage in a composite. These two sources are:

1. Fabrication or processing defects and

2. In-field or service defects

The defects in these two categories are listed below.

1) *Fabrication or Processing Defects*:

The defects that can occur during fabrication or processing are listed below:

1. Abrasions, scratches, dents and punctures

2. Cut fibres

3. Knots and kinks in fibres

4. Improper splicing (joining) of layers

5. Voids (due to poor processing, high humidity)

6. Inferior quality of the materials used

7. Improper curing of resin

8. Resin rich or resin lean areas due to improper distribution of resin

9. Inclusions and contamination

10. Mandrel removal problem

11. Machining problems

12. Improper tooling

13. Tool drop causing low energy impact which results in impact damage

2) *In-field or Service Defects*:

The defects that can occur during in-field or service are listed below:

1. Shock

2. Environmental cycle of temperature and humidity

3. Exposure to hazardous chemicals

4. Exposure to radiations

5. Bacterial degradation

6. Vibrations

7. Improper handling and storage

8. Tool drop

9. Abrasions, dents and punctures

10. Corrosion

11. Erosion due to sand and dust

12. Improper maintenance or repair

Damage Mechanics

Damage mechanics is concerned with the representation, or modeling, of damage of materials that is suitable for making engineering predictions about the initiation, propagation, and fracture of materials without resorting to a microscopic description that would be too complex for practical engineering analysis. Damage mechanics illustrates the typical engineering approach to model complex phenomena. To quote Dusan Krajcinovic, "It is often argued that the ultimate task of engineering research is to provide not so much a better insight into the examined phenomenon but to supply a rational predictive tool applicable in design". Damage mechanics is a topic of applied mechanics that relies heavily on continuum mechanics. Most of the work on damage mechanics uses state variables to represent the *effects* of damage on the stiffness and remaining life of the material that is damaging as a result of thermomechanical load and ageing. The state variables may be measurable, e.g., crack density, or inferred from the *effect* they have on some macroscopic property, such as stiffness, coefficient of thermal expansion, remaining life, etc. The state variables have conjugate thermodynamic forces that motivate further damage. Initially the material is pristine, or *intact*. A damage activation criterion is needed to predict damage initiation. Damage evolution does not progresses spontaneously after initiation, thus requiring a damage evolution model. In plasticity like formulations, the damage evolution is controlled by a hardening function but this requires additional phenomenological parameters that must be found through experimentation, which is expensive, time consuming, and virtually no one does. On the other hand, micromechanics of damage formulations are able to predict both damage initiation and evolution without additional material properties.

Damage Mechanisms in Fibrous Composites:

The damage mechanisms in a fibrous composite are broadly categorized as:

1. Micro-level damage mechanisms

2. Macro-level damage mechanisms and

3. Coupled micro-macro-level damage mechanisms

The local level mechanisms are further subcategorized based on constituent level as

i. Fibre level damage mechanisms

ii. Matrix level damage mechanisms and

iii. Coupled fibre-matrix level damage mechanisms

A. *Micro-level Damage Mechanisms*:

First, we will look at the micro-level mechanisms in detail as follows:

a) Fibre Level Damage Mechanisms:

The fibre failure mode is considered to be the most catastrophic mode of failure in laminates. This is because the fibre is the load carrying constituent. The failure of fibres can take place due to various stress components. The damage mechanisms for fibre are explained below in detail.

1) *Fibre Fracture/Breaking*:

The fibre breaks into two or more pieces along its length when the axial tensile stress (or strain) in the fibre exceeds the axial strength (or maximum allowable strain) of the fibre. This kind of fracture occurs in brittle fibres. Such fractures are more catastrophic in nature than other modes of fibre failure.

The fibre fracture may also take place in shearing when the shear stress or strain exceeds the maximum allowable stress or strain.

The fibre fracture is depicted in Figure (a).

2) Fibre Buckling or Kinking:

This type of failure occurs when the axial load on the fibre is compressive in nature. The axial compressive stress causes the fibre to buckle. This form of fibre failure is also called as fibre kinking. The critical stress at which the kinking takes place is function of material properties of fibre and matrix properties and the distribution of fibres in the matrix. In general, the fibre kinking first starts at the site of fibre misalignment or local defects.

It is seen that the kinking of fibres takes place in a sharply defined region. This region is called as kink band. In general, the kink band is oriented at an angle with respect to fibre direction.

This mechanism is one of the key failure mechanisms for laminates under compression. This failure mechanism triggers the other failure mechanisms leading to a complex and inter-related mechanisms.

The fibre kinking is depicted in Figure (b).

3) Fibre Bending:

The bending of fibre can take place under flexural load. The bending of fibres also depends upon the properties of fibre and matrix along with the fibre arrangement.

The fibre bending is shown in Figure (c).

4) Fibre Splitting:

The fibre fails in this mode when the transverse or hoop stresses in the fibre exceeds the maximum allowable value. Further, this can also happen when these stresses in the interface/interphase region (region in matrix very close to the fibre) exceed the maximum allowable stress. The fibre splitting is elucidated in Figure (d).

5) Fibre Radial Cracking:

The hoop stresses can also cause the radial cracking of the fibre. This type of cracking is seen in some of the fibres. The radial cracking of a fibre is shown in Figure (e).

b) Matrix Level Damage Mechanisms:

There are two main damage mechanisms in matrix. These are: Matrix cracking and fibre interfacial debonding. These are explained below.

1) Matrix Cracking:

When the stress in the matrix exceeds the strength of the matrix, matrix cracks are developed. There are two types of matrix cracks that are developed in a unidirectional lamina. The cracks are either perpendicular or parallel to the fibre direction. In the first type, the cracks are developed when axial stress in the lamina is tensile in nature. In the second type, the cracks are developed when the in-plane transverse stress in the lamina is tensile in nature.

It is generally seen that the matrix cracks develop along the preferred directions in unidirectional lamina. The matrix cracks which are parallel to the fibre direction cause significant modulus degradation whereas the matrix cracks which are perpendicular to the fibre direction cause less degradation in modulus. The first mode of damage is very critical as one of them causes significant degradation. The second mode can go undetected sometimes. This is very dangerous from safety point of view. For example, for gas pipes leakage is an important criterion. If such damage is not detectable, it can lead to a catastrophe. This damage is shown in Figure (a), (b).

2) Fibre Interfacial Cracking:

When the in-plane transverse stresses in matrix are tensile in nature, the weaker interface between fibre and matrix is broken. A crack in the matrix region at this location is initiated. This crack grows along the fibre length. This leads to the debonding of the interphase between fibre and matrix. This mode of damage is also called "transverse fibre debonding". This damage is shown in Figure (c).

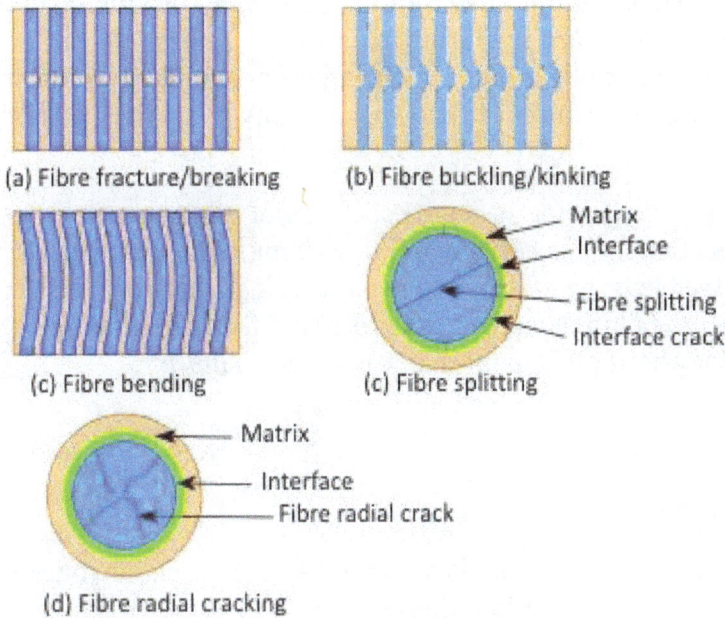

(a) Fibre fracture/breaking   (b) Fibre buckling/kinking

(c) Fibre bending   (c) Fibre splitting

(d) Fibre radial cracking

Fibre-level damage mechanisms

Failure Cause

Failure causes are defects in design, process, quality, or part application, which are the underlying cause of a failure or which initiate a process which leads to failure. Where failure depends on the user of the product or process, then human error must be considered.

Component Failure / Failure Modes

A part failure mode is the way in which a component fails "functionally" on component level. Often a part has only a few failure modes. Thus a relay may fail to open or close contacts on demand. The failure mechanism that caused this can be of many different kinds, and often multiple factors play a role at the same time. They include corrosion, welding of contacts due to an abnormal electric current, return spring fatigue failure, unintended command failure, dust accumulation and blockage of mechanism, etc. Seldom only one cause (hazard) can be identified that creates system failures. The real root causes can in theory in most cases be traced back to some kind of human error, e.g. design failure, operational errors, management failures, maintenance induced failures, specification failures, etc.

Failure Scenario

A scenario is the complete identified possible sequence and combination of events, failures (failure modes), conditions, system states, leading to an end (failure) system state. It starts from causes (if known) leading to one particular end effect (the system failure condition). A failure scenario is for a system the same as the failure mechanism is for a component. Both result in a failure mode (state) of the system / component.

Rather than the simple description of symptoms that many product users or process participants might use, the term failure scenario / mechanism refers to a rather complete description, including the preconditions under which failure occurs, how the thing was being used, proximate and ultimate/final causes (if known), and any subsidiary or resulting failures that result.

The term is part of the engineering lexicon, especially of engineers working to test and debug products or processes. Carefully observing and describing failure conditions, identifying whether failures are reproducible or transient, and hypothesizing what combination of conditions and sequence of events led to failure is part of the process of fixing design flaws or improving future iterations. The term may be applied to mechanical systems failure.

Types of Failure Causes

Mechanical Failure

Some types of mechanical failure mechanisms are: excessive deflection, buckling, ductile fracture, brittle fracture, impact, creep, relaxation, thermal shock, wear, corrosion, stress corrosion cracking, and various types of fatigue. Each produces a different type of fracture surface, and other indicators near the fracture surface(s). The way the product is loaded, and the loading history are also important factors which determine the outcome. Of critical importance is design geometry because stress concentrations can magnify the applied load locally to very high levels, and from which cracks usually grow.

Over time, as more is understood about a failure, the failure cause evolves from a description of symptoms and outcomes (that is, effects) to a systematic and relatively abstract model of how, when, and why the failure comes about (that is, causes).

The more complex the product or situation, the more necessary a good understanding of its failure cause is to ensuring its proper operation (or repair). Cascading failures, for example, are particularly complex failure causes. Edge cases and corner cases are situations in which complex, unexpected, and difficult-to-debug problems often occur.

Failure by Corrosion

Materials can be degraded by their environment by corrosion processes, such as rusting in the case of iron and steel. Such processes can also be affected by load in the mechanisms of stress corrosion cracking and environmental stress cracking.

Failure Mechanisms

1) Fibre Pullout:

The fibre pullout takes place when the bonding between fibre and matrix is weakened and the fibres are subjected to tensile stresses. If the fibres are already broken then the fibres just slide through the matrix and come out of it. This phenomenon is called fibre pullout.

The fibre pullout is shown in Figure (a).

2) Fibre Breakage and Interfacial Debonding:

When the fibres break the interface close to the tip of broken fibre, acts as a site of stress concentration. The interface may then fail, leading to debonding of the fibre from matrix.

The fibre breakage leading to interfacial debonding is shown in Figure (b).

3) Transverse Matrix Cracking:

The interface failure causing debonding (as in fibre breaking and interfacial debonding in above case) from the matrix may act like as a stress concentration site for the in-plane transverse tensile stress. When this stress exceeds the limiting stress in matrix, it leads to through thickness transverse crack in the matrix.

The through thickness transverse matrix cracking is shown in Figure (c).

4) Fibre Failure due to Matrix Cracking:

The matrix cracks formed (as in matrix cracking case above) may terminate at fibre interface at low strains, while, at high strains, the stress at the crack tip may exceed the fracture stress of the fibres, leading fibre failure.

The fibre failure due to matrix cracking is depicted in Figure (d).

5) Interfacial Shear Failure:

The fibre fracture or fibre failure due to matrix cracking may cause the matrix crack to propagate as macro-crack under opening mode until it hits an interface. The shear stresses may cause its propagation in sliding mode leading to a progressive failure of the interface.

The interfacial shear failure is shown in Figure (e).

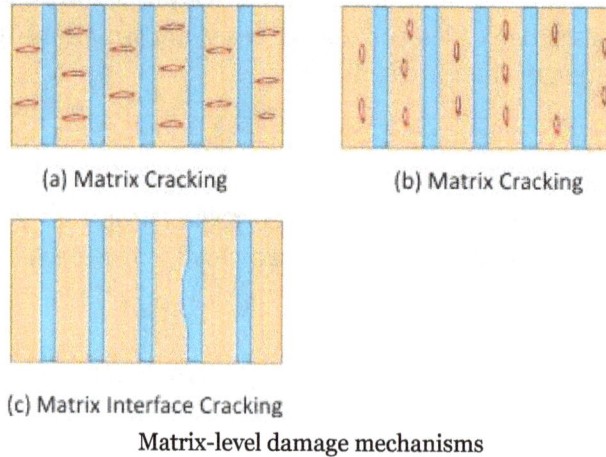

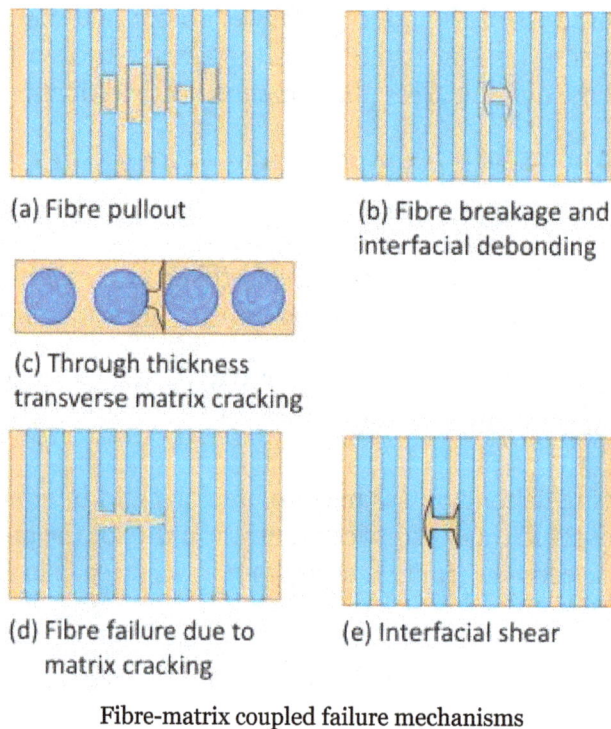

(a) Matrix Cracking   (b) Matrix Cracking

(c) Matrix Interface Cracking

Matrix-level damage mechanisms

(a) Fibre pullout   (b) Fibre breakage and
                              interfacial debonding

(c) Through thickness
transverse matrix cracking

(d) Fibre failure due to   (e) Interfacial shear
matrix cracking

Fibre-matrix coupled failure mechanisms

Macro-level Failure Mechanism

The macro-level mechanisms are laminate level mechanisms. Here, we are addressing the delamination. It is seen that the adjacent layers are bonded together by a thin layer of resin between them. This interface layer transfers the displacement and force from one layer to another layer. When this interface layer weakens or damages completely, it causes the adjacent layers to separate. This mode of failure is called delamination. It is shown in Figure.

Delamination reduces the strength and stiffness and thus limits the life of a structure. Further, it causes stress concentration in load bearing plies and a local instability leading to a further growth of delamination which results in a compressive failure of the laminate. In these two cases delamination leads to a redistribution of structural load paths which, in turn, precipitates structural failure. Hence, delamination indirectly affects the final failure of the structure thus affecting its life. Therefore, delamination is known as the most prevalent life limiting damage growth mode.

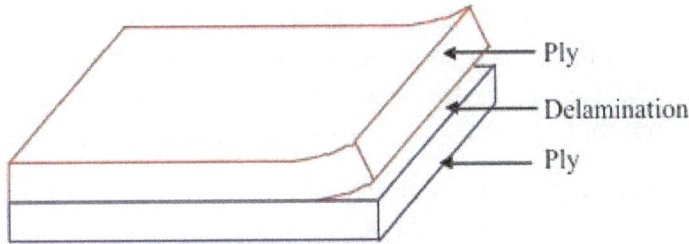

Macro-level damage mechanism (Delamination)

Causes of Delamination

Delamination can occur due to variety of reasons. The situations which can lead to delamination initiation and its growth are explained below.

a) Manufacturing Defects

This is the most common reason for existence of delaminations in a laminate. Improper laying of laminae, insufficient curing temperature; pressure and duration of curing, air pockets and inclusions are some of the reasons which lead the manufacturing defects causing delamination.

b) Loading Generating Transverse Stresses

The interface is weaker in transverse strength as compared to the layers. Hence, its failure is dominated by the transverse stresses. The interface generally fails under tensile load applied normal to it. Also, the delamination can take place due to compressive stresses in its inplane direction causing buckling, which in turn, causes delamination.

The inplane loads applied to angle ply laminate can cause delamination in it. This is because the bending-stretching coupling can give rise to transverse stresses in the interface. A schematic illustration of how axial tensile loading of angle ply laminates cause rotation of the plies is shown in Figure (b). This rotation of the plies generates the interlaminar shear stresses, which is one of the crucial factors in delamination.

Note: The Inter-laminar stresses are the stresses in the interface between two adjacent layers. The existence to these stresses is shown in various references. Further, these stresses can be very high locally depending upon various situations.

c) Laminate Geometry

The geometry of the laminate can lead to a three dimensional state of stress locally in the interface leading to high interlaminar stresses. Some of the geometries of the laminate and structures are shown below in which delamination damage will be a major damage mode.

i.   Free Edge:

The free edges of the laminate have very high transverse normal (σ_{zz}) and shear (τ_{xz}, τ_{yz}) stresses. It is shown that significant interlaminar stresses are induced in regions near the laminate free edges. Interlaminar stresses near the free edges can be controlled to an extent through the choice of materials, fibre orientations, stacking sequence, layer thickness and the use of functionally graded materials. However, when free edges are present, interlaminar stresses can be completely eliminated through the use of a homogeneous material, locally.

The delamination shown in Figure, infact, is an edge delamination.

ii.   Notch:

Notch in the laminates acts like an external crack giving rise to high three dimensional stress state in the vicinity of the notch.

iii. Cut-out:

Cutouts are inevitable in structures. Cutouts are made to pass electric wires; fluid passage as in the wings, doors and windows in the fuselage of an air vehicle. These are, especially in aerospace vehicles, made also to reduce the weight of the component. The cutout boundaries act like free edges leading to significant transverse stresses. This is one of the most common site for onset of delamination. A laminate with cutout is shown in Figure (d).

iv.   Ply Drop/Termination:

The optimum design of composite structures in air vehicles is important. As a result of the optimization (e.g. weight minimization) process or sometimes purely due to geometric requirements/constraints, one or more of the plies have to be terminated (also known as "ply drop") inside the laminate. The region of ply termination acts like a region of high stresses for neighbouring laminae which can be a reason for delamination of the plies adjacent to the ply drop region. A ply drop in laminate is shown in Figure (e).

v.   Bonded Joints:

Sometimes laminates are bonded together using resin. Improper bonding leads to weaker joints. When such weak joints are subjected to serve loading conditions delamination can occur. A bonded joint in composite is shown in Figure (f).

vi.   Bolted Joints:

Sometimes it is required to attach the composite structures to metallic structures. In such situations, bolted joints are imperative. The free edges of the cutout made in composite and additional load applied by tightening of the joint leads to a complex local state of stress. When these composite structures are T or L sections carrying additional loads, the situation is the worst. In such a situation delamination starts at cutout edges or at the curved edges of the T or L sections. A L-bolted joint is shown in Figure (g).

vii.   Doublers:

These are needed due to geometric or functional requirements in the structures. In this case a laminate is split into two or more set of laminae (or vice a versa). Thus, at the bifurcation laminae (or where the laminae join together to form laminate) give rise to high stresses. These locations are potential zones for delamination initiation. Typical doublers are shown in Figure (h).

Suppression of Delamination

Several possible design changes are suggested for delaying/suppressing the onset and growth of delamination.

The primary cause of delamination is the low interlaminar fracture toughness. This is due to brittle nature of most resins (epoxy) used as matrix material, which have low mode I fracture toughness. The suggested models for improving this property are:

a. Adding thermoplastics, interleafing soft and hard layers, increasing length of cross-links

b. Adding second phase materials to matrix like rubber; chopped fibre, fibrils, etc.

c. Through thickness reinforcement by 3D braiding or stitching

Coupled Micro-Macrolevel Failure Mechanisms

The transverse matrix cracking of a lamina as shown in Figure (c) is an important failure mechanism. The through thickness transverse crack may propagate to neighbouring lamina causing it to break.

There can be another scenario that this crack terminates at the neighboring interface. This crack front act as a stress concentration site for interface between the adjacent layer causing it to weaken, thus initiating a delamination crack in the interface. This delamination growth can lead to failure of the laminate. This is depicted in Figure (a).

A third scenario is also possible in which the transverse through thickness crack leads to interface crack in adjacent layer causing partial delamination. This delamination may cause a transverse crack in the next layer. Then this crack initiates a interfacial debonding of that layer and so on causing the failure of laminate.

The coupling between the transverse cracking of lamina and delamination is depicted in Figure (b).

Thus, the transverse cracking of lamina and delamination are strongly coupled.

(a) Matrix cracking
leading to delamination

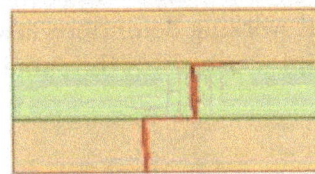

(b) Alternate matrix
cracking delamination

Coupled micro-macro damage mechanisms

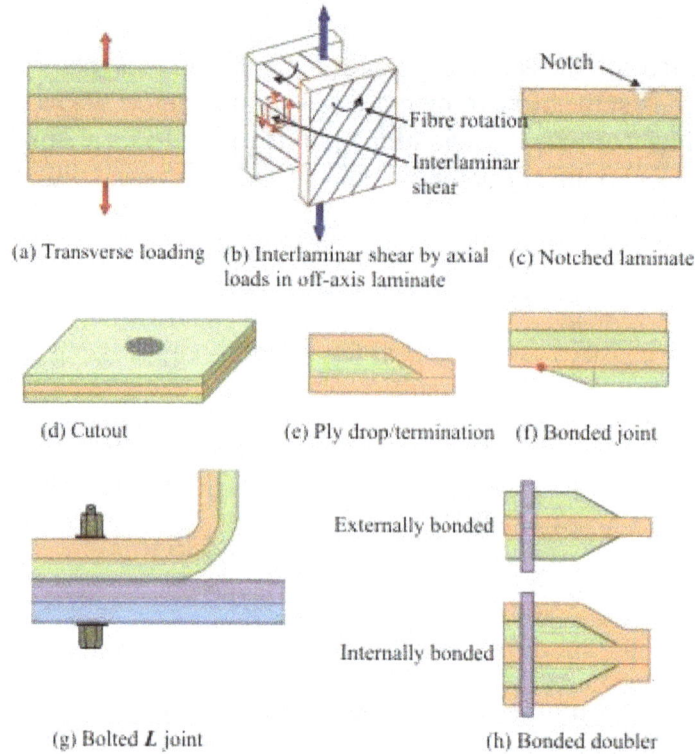

(a) Transverse loading   (b) Interlaminar shear by axial   (c) Notched laminate
                          loads in off-axis laminate

(d) Cutout   (e) Ply drop/termination   (f) Bonded joint

Externally bonded

Internally bonded

(g) Bolted *L* joint   (h) Bonded doubler

Situations conducive for delamination

Material Failure Theory

Failure theory is the science of predicting the conditions under which solid materials fail under the action of external loads. The failure of a material is usually classified into brittle failure (fracture) or ductile failure (yield). Depending on the conditions (such as temperature, state of stress, loading rate) most materials can fail in a brittle or ductile manner or both. However, for most practical situations, a material may be classified as either brittle or ductile. Though failure theory has been in development for over 200 years, its level of acceptability is yet to reach that of continuum mechanics.

In mathematical terms, failure theory is expressed in the form of various failure criteria which are valid for specific materials. Failure criteria are functions in stress or strain space which separate "failed" states from "unfailed" states. A precise physical definition of a "failed" state is not easily quantified and several working definitions are in use in the engineering community. Quite often, phenomenological failure criteria of the same form are used to predict brittle failure and ductile yield.

Material Failure

In materials science, material failure is the loss of load carrying capacity of a material unit. This definition *per se* introduces the fact that material failure can be examined in different scales, from microscopic, to macroscopic. In structural problems, where the structural response may be beyond the initiation of nonlinear material behaviour, material failure is of profound importance for the determination of the integrity of the structure. On the other hand, due to the lack of globally accepted fracture criteria, the determination of the structure's damage, due to material failure, is still under intensive research.

Types of Material Failure

Material failure can be distinguished in two broader categories depending on the scale in which the material is examined:

Microscopic Failure

Microscopic material failure is defined in terms of crack propagation and initiation. Such methodologies are useful for gaining insight in the cracking of specimens and simple structures under well defined global load distributions. Microscopic failure considers the initiation and propagation of a crack. Failure criteria in this case are related to microscopic fracture. Some of the most popular failure models in this area are the micromechanical failure models, which combine the advantages of continuum mechanics and classical fracture mechanics. Such models are based on the concept that during plastic deformation, microvoids nucleate and grow until a local plastic neck or fracture of the intervoid matrix occurs, which causes the coalescence of neighbouring voids. Such a model, proposed by Gurson and extended by Tvergaard and Needleman, is known as GTN. Another approach, proposed by Rousselier, is based on continuum damage mechanics (CDM) and thermodynamics. Both models form a modification of the von Mises yield potential by introducing a scalar damage quantity, which represents the void volume fraction of cavities, the porosity f.

Macroscopic failure

Macroscopic material failure is defined in terms of load carrying capacity or energy storage capacity, equivalently. Li presents a classification of macroscopic failure criteria in four categories:

- Stress or strain failure

- Energy type failure (S-criterion, T-criterion)

- Damage failure

- Empirical failure.

Five general levels are considered, at which the meaning of deformation and failure is interpreted differently: the structural element scale, the macroscopic scale where macroscopic stress and strain are defined, the mesoscale which is represented by a typical void, the microscale and the atomic scale. The material behavior at one level is considered as a collective of its behavior at a sub-level. An efficient deformation and failure model should be consistent at every level.

Brittle Material Failure Criteria

Failure of brittle materials can be determined using several approaches:

- Phenomenological failure criteria

- Linear elastic fracture mechanics

- Elastic-plastic fracture mechanics

- Energy-based methods

- Cohesive zone methods

Phenomenological Failure Criteria

The failure criteria that were developed for brittle solids were the maximum stress/strain criteria. The maximum stress criterion assumes that a material fails when the maximum principal stress σ_1 in a material element exceeds the uniaxial tensile strength of the material. Alternatively, the material will fail if the minimum principal stress σ_3 is less than the uniaxial compressive strength of the material. If the uniaxial tensile strength of the material is σ_t and the uniaxial compressive strength is σ_c, then the safe region for the material is assumed to be:

$$\sigma_c < \sigma_3 < \sigma_1 < \sigma_t$$

Note that the convention that tension is positive has been used in the above expression.

The maximum strain criterion has a similar form except that the principal strains are compared with experimentally determined uniaxial strains at failure, i.e.,

$$\varepsilon_c < \varepsilon_3 < \varepsilon_1 < \varepsilon_t$$

The maximum principal stress and strain criteria continue to be widely used in spite of severe shortcomings.

Numerous other phenomenological failure criteria can be found in the engineering literature. The degree of success of these criteria in predicting failure has been limited. For brittle materials, some popular failure criteria are:

- criteria based on invariants of the Cauchy stress tensor

- the Tresca or maximum shear stress failure criterion

- the von Mises or maximum elastic distortional energy criterion

- the Mohr-Coulomb failure criterion for cohesive-frictional solids

- the Drucker-Prager failure criterion for pressure-dependent solids

- the Bresler-Pister failure criterion for concrete

- the Willam-Warnke failure criterion for concrete

- the Hankinson criterion, an empirical failure criterion that is used for orthotropic materials such as wood.

- the Hill yield criteria for anisotropic solids

- the Tsai-Wu failure criterion for anisotropic composites

- the Johnson–Holmquist damage model for high-rate deformations of isotropic solids

- the Hoek-Brown failure criterion for rock masses

- the Cam-Clay failure theory for soils

Linear Elastic Fracture Mechanics

The approach taken in linear elastic fracture mechanics is to estimate the amount of energy needed to grow a preexisting crack in a brittle material. The earliest fracture mechanics approach for unstable crack growth is Griffiths' theory. When applied to the mode I opening of a crack, Griffiths' theory predicts that the critical stress (σ) needed to propagate the crack is given by

$$\sigma = \sqrt{\frac{2E\gamma}{\pi a}}$$

where E is the Young's modulus of the material, γ is the surface energy per unit area of the crack, and a is the crack length for edge cracks or $2a$ is the crack length for plane cracks. The quantity $\sigma\sqrt{\pi a}$ is postulated as a material parameter called the fracture toughness. The mode I fracture toughness for plane strain is defined as

$$K_{\text{Ic}} = Y\sigma_c\sqrt{\pi a}$$

where σ_c is a critical value of the far field stress and Y is a dimensionless factor that depends on the geometry, material properties, and loading condition. The quantity K_{Ic} is related to the stress intensity factor and is determined experimentally. Similar quantities K_{IIc} and K_{IIIc} can be determined for mode II and model III loading conditions.

The state of stress around cracks of various shapes can be expressed in terms of their stress intensity factors. Linear elastic fracture mechanics predicts that a crack will extend when the stress intensity factor at the crack tip is greater than the fracture toughness of the material. Therefore, the critical applied stress can also be determined once the stress intensity factor at a crack tip is known.

Energy-based Methods

The linear elastic fracture mechanics method is difficult to apply for anisotropic materials (such as composites) or for situations where the loading or the geometry are complex. The strain energy release rate approach has proved quite useful for such situations. The strain energy release rate for a mode I crack which runs through the thickness of a plate is defined as

$$G_I := \frac{P}{2t}\frac{du}{da}$$

where P is the applied load, t is the thickness of the plate, u is the displacement at the point of application of the load due to crack growth, and a is the crack length for edge cracks or $2a$ is the crack length for plane cracks. The crack is expected to propagate when the strain energy release rate exceeds a critical value G_{Ic} - called the critical strain energy release rate.

The fracture toughness and the critical strain energy release rate for plane stress are related by

$$G_{\text{Ic}} = \frac{1}{E} K_{\text{Ic}}^2$$

where E is the Young's modulus. If an initial crack size is known, then a critical stress can be determined using the strain energy release rate criterion.

Ductile Material Failure Criteria

Criteria used to predict the failure of ductile materials are usually called yield criteria. Commonly used failure criteria for ductile materials are:

- the Tresca or maximum shear stress criterion.
- the von Mises yield criterion or distortional strain energy density criterion.
- the Gurson yield criterion for pressure-dependent metals.
- the Hosford yield criterion for metals.
- the Hill yield criteria.
- various criteria based on the invariants of the Cauchy stress tensor.

The yield surface of a ductile material usually changes as the material experiences increased deformation. Models for the evolution of the yield surface with increasing strain, temperature, and strain rate are used in conjunction with the above failure criteria for isotropic hardening, kinematic hardening, and viscoplasticity. Some such models are:

- the Johnson-Cook model
- the Steinberg-Guinan model
- the Zerilli-Armstrong model
- the Mechanical threshold stress model
- the Preston-Tonks-Wallace model

There is another important aspect to ductile materials - the prediction of the ultimate failure strength of a ductile material. Several models for predicting the ultimate strength have been used by the engineering community with varying levels of success. For metals, such failure criteria are usually expressed in terms of a combination of porosity and strain to failure or in terms of a damage parameter.

Macroscopic Failure Theories

In the following we will present some of the popular macroscopic failure theories used in design and analysis of composites.

In these theories we will use following quantities and symbols.

1. X denotes the ultimate normal stress magnitude in fibre direction (1-direction).

2. Y denotes the ultimate normal stress magnitude in in-plane transverse direction (2-direction).

3. Z denotes the ultimate normal stress magnitude in transverse direction (3-direction)

4. Subscript T and C denote tension and compression, respectively.

5. Q, R and S denote the ultimate shear stresses corresponding to 23, 13 and 12 planes.

We will see some definitions related to failure theories.

Strength Ratio (SR):

It is defined as the ratio of maximum load which can be applied such that a lamina does not fail to the actual load applied. Thus,

$$SR = \frac{\text{Maximum load which can be applied}}{\text{Actual load applied}}$$

(1)

This concept can be extended to any failure theory. The strength ratio gives the factor by which the actual applied load can be increased or decreased upto a lamina failure. For example, if $SR > 1$, it means that the lamina is safe and load applied can be increased by this factor and if $SR < 1$, it means that the lamina is unsafe and the load applied must be decreased by this factor. It is needless to say that when $SR = 1$ the condition corresponds to failure load.

Failure Envelope:

The failure envelope is a surface formed by various combinations of normal and shear stresses (or strains) that can be applied to a lamina just before it fails. Thus, any state of stress (or strain) which lies inside the envelope is safe whereas the one which lies on or outside the envelope is unsafe.

1. Maximum Stress Theory:

This theory is a direct extension of maximum normal stress theory proposed by Rankiene and maximum stress theory proposed by Tresca for homogeneous, isotropic materials. In this theory the three normal and three shear stress components are compared with corresponding ultimate stresses. A given normal stress is compared with corresponding positive and negative, that is tensile and compressive ultimate stresses. The magnitude of shear stress is compared with corresponding ultimate shear stress.

Thus, the maximum stress theory results in the following expression for the safe condition.

For normal stresses,

$$X_c < \sigma_1 < X_T$$
$$Y_c < \sigma_2 < Y_T$$
$$Z_c < \sigma_3 < Z_T \qquad (2)$$

For shear stresses,

$$|\tau_{23}| < Q \qquad |\sigma_4| < Q$$
$$|\tau_{13}| < R \qquad |\sigma_5| < R$$
$$|\tau_{12}| < S \quad \text{OR} \quad |\sigma_6| < S \qquad (3)$$

Thus, according to this theory initiation of failure will correspond to one or more inequalities in Equations (2) and (3) become an equality. The maximum stress theory can be represented as intersecting planes in $3D$ stress space or intersecting lines in $2D$ stress space.

The $\sigma_1 - \sigma_2 - \sigma_3$ stress space is shown as intersecting planes in Figure. The region inside this space is regarded as safe, whereas any point on or outside the intersecting planes will be an unsafe or a failure point. A safe state of stress with normal stresses alone is shown inside the envelope.

A fully 3D state of stress will represent an envelope or surface in six dimensional stress space.

The maximum stress theory for planar state of stress is given for normal stresses as

$$X_c < \sigma_1 < X_T$$
$$Y_c < \sigma_2 < Y_T \qquad (4)$$

and for shear stress as

$$|\tau_{12}| < s \qquad (5)$$

Now, consider that an off axis lamina is subjected to an axial stress of σ_{xx}. Then, we can write the maximum stress theory for the planar state of stress for off axis lamina as follows.

Recalling the stress transformation for planar state of stress, we write the stress components in principal material directions as

$$\sigma_1 = \sigma_{xx} \cos^2 \theta$$
$$\sigma_2 = \sigma_{xx} \sin^2 \theta$$
$$\tau_{12} = -\sigma_{xx} \sin \theta \cos \theta \qquad (6)$$

Thus, the maximum stress theory for off-axis lamina loaded axially can be written as

$$X_C < \sigma_{xx} \cos^2 \theta < X_T$$
$$Y_C < \sigma_{xx} \sin^2 \theta < Y_T$$
$$|-\sigma_{xx} \sin \theta \cos \theta| < S \qquad (7)$$

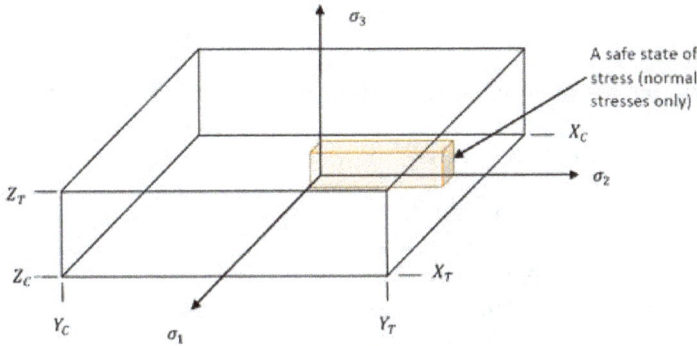

Failure envelope for normal stress space with an example safe stress state inside the envelope

2. Maximum Strain Theory:

Maximum strain theory is equivalent to maximum stress theory. This theory is based on maximum normal strain theory of St. Venant and the strain equivalent of maximum shear stress theory of Tresca for isotropic materials.

According to this theory, a lamina fails if either of the normal strain exceeds the maximum allowable strain in tension or compression or any of the shear strain exceeds the maximum allowable shear strain. The inequalities resulting are:

For normal strains,

$$\varepsilon_1^C < \varepsilon_1 < \varepsilon_1^T$$
$$\varepsilon_2^C < \varepsilon_2 < \varepsilon_2^T$$
$$\varepsilon_3^C < \varepsilon_3 < \varepsilon_3^T \tag{8}$$

For shear stresses,

$$
\begin{array}{ll}
|\gamma_{23}| < \Gamma_{23} & |\varepsilon_4| < \Gamma_{23} \\
|\gamma_{13}| < \Gamma_{13} & |\varepsilon_5| < \Gamma_{13} \\
|\gamma_{12}| < \Gamma_{12} \quad \text{OR} & |\varepsilon_6| < \Gamma_{12} \tag{9}
\end{array}
$$

where, $\varepsilon_1^T, \varepsilon_2^T, \varepsilon_2^T$ and $\varepsilon_1^C, \varepsilon_2^C, \varepsilon_2^C$ are the ultimate normal strains in tension and compression, respectively. Further, $\Gamma_{23}, \Gamma_{13}, \Gamma_{12}$ are ultimate shear strains in 23, 13, 12 planes, respectively.

Thus, according to this theory initiation of failure will correspond to one or more inequalities in Equations (8) and (9) become equality.

The maximum strain theory for planar stress can be expressed as:

$$\varepsilon_1^C < \varepsilon_1 < \varepsilon_1^T$$
$$\varepsilon_2^C < \varepsilon_2 < \varepsilon_2^T \tag{10}$$

and for shear strain as:

$$|\gamma_{12}| < \Gamma_{12} \tag{11}$$

The strains can be obtained from constitutive equation for strains in terms stresses as:

$$\varepsilon_1 = \sigma_1 - \nu_{12}\,\sigma_2$$
$$\varepsilon_2 = \sigma_2 - \nu_{21}\,\sigma_1$$
$$\gamma_{12} = \frac{\tau_{12}}{G_{12}} \tag{12}$$

These equations can be put in Equation (10) and Equation. (11). Further, for axial stress applied σ_{xx} we can write the stresses in principal directions as in Equation (6).

Note: The maximum stress and maximum strain theories are similar. In both theories there is no interaction between various components of stress or strain. However, the two theories yield different results.

3. Tsai-Hill Theory:

This theory is an extension of distortional energy yield criterion of von-Mises for isotropic materials as applied to anisotropic materials. It is known that total strain energy in a body is composed of two parts: One is distortion energy which cause change in shape and second one is dilation energy which causes the change in size or volume. In the von-Mises criterion it is assumed that the material fails when the maximum distortion energy of the body exceeds the distortion energy corresponding to yielding of the same material in tension.

Hill extended the von-Mises distortion energy criterion of isotropic materials to anisotropic materials. Later Tsai extended this criterion for anisotropic materials to a unidirectional lamina. Hence, the theory is called Tsai-Hill theory.

According to this theory the failure takes place when the stress state is such that:

$$f\left(\sigma_{ij}\right) = F\left(\sigma_2 - \sigma_3\right)^2 + G\left(\sigma_3 - \sigma_1\right)^2 + H\left(\sigma_1 - \sigma_2\right)^2 + 2L\,\sigma_4^2 + 2M\,\sigma_S^2 + 2N\,\sigma_6^2 = 1 \tag{13}$$

which upon simplifications can be written as:

$$\left(G + H\right)\sigma_1^2 + \left(F + H\right)\sigma_2^2 + \left(F + G\right)\sigma_3^2 - 2H\,\sigma_1\,\sigma_2 - 2G\,\sigma_1\,\sigma_3 - 2F\,\sigma_2\,\sigma_3$$
$$+ 2L\,\sigma_4^2 + 2M\,\sigma_S^2 + 2N\,\sigma_6^2 = 1 \tag{14}$$

where, *F, G, H, L, M* and *N* are the material strength parameters. Thus, any state of stress which lies inside this envelope is safe and the one which lies on or outside the envelope is unsafe.

The strength parameters correspond to failure stresses in one dimensional loading. These can be obtained by a set of thought experiments. For example, consider that for the pure shear loading in 2-3 plane, that is with $\sigma_4 = \tau_{23} \neq 0$, with corresponding shear strength Q and all other stress components are zero, the Equation (14) becomes

$$2L = \frac{1}{Q^2} \tag{15}$$

Similarly, for the other two shear stress components, we can get

$$2M = \frac{1}{R^2}$$

$$2N = \frac{1}{S^2} \tag{16}$$

Now the strength parameters F, G and H are obtained by three states of stress. The state of stress $\sigma_1 \neq 0$ with corresponding strength X and all other stress components being zero, in Equation (14) leads to

$$G + H = \frac{1}{X^2} \tag{17}$$

Now the conditions $\sigma_2 \neq 0$ and $\sigma_3 \neq 0$ (and other stress components being zero) in Equation (14) result in

$$F + H = \frac{1}{Y^2}$$

$$F + G = \frac{1}{Z^2} \tag{18}$$

solving the simultaneous equations in Equation (17) and Equation (18), we get the required strength parameters as

$$2H = \frac{1}{X^2} + \frac{1}{Y^2} - \frac{1}{Z^2}$$

$$2G = \frac{1}{X^2} - \frac{1}{Y^2} + \frac{1}{Z^2}$$

$$2F = -\frac{1}{X^2} + \frac{1}{Y^2} + \frac{1}{Z^2} \tag{19}$$

Thus, Equation (14) becomes

$$\frac{\sigma_1^2}{X^2} + \frac{\sigma_2^2}{Y^2} + \frac{\sigma_3^2}{Z^2} - \sigma_1 \sigma_2 \left(\frac{1}{X^2} + \frac{1}{Y^2} - \frac{1}{Z^2} \right) - \sigma_1 \sigma_3 \left(\frac{1}{X^2} - \frac{1}{Y^2} + \frac{1}{Z^2} \right)$$

$$- \sigma_2 \sigma_3 \left(-\frac{1}{X^2} + \frac{1}{Y^2} + \frac{1}{Z^2} \right) + \frac{\sigma_4^2}{Q^2} + \frac{\sigma_5^2}{R^2} + \frac{\sigma_6^2}{S^2} = 1 \tag{20}$$

This is Tsai-Hill theory for 3D state of stress. Note that this is quadratic in stress terms with no linear terms.

Now consider a transversely isotropic material with inplane stresses as the significant stresses. For this planar state of stress, we have $\sigma_3 = \sigma_4 = \sigma_5 = 0$ and remaining stress components are non zero. In this case the failure envelope becomes a three dimensional space. Thus, the failure condition in Equation (14) becomes

$$(G+H)\sigma_1^2 + (F+H)\sigma_2^2 - 2H\sigma_1\sigma_2 + 2N\sigma_6^2 = 1 \quad (21)$$

Now, using the strength parameters from Equation (17), Equation (18) and Equation (19), we get

$$\frac{\sigma_1^2}{X^2} + \frac{\sigma_2^2}{Y^2} - \sigma_1\sigma_2\left(\frac{1}{X^2} + \frac{1}{Y^2} - \frac{1}{Z^2}\right) + \frac{\sigma_6^2}{S^2} = 1 \quad (22)$$

For transverse isotropy, we also have $Y = Z$. Thus, the above equation is rearranged as

$$\frac{\sigma_1^2}{X^2} - \frac{\sigma_1\sigma_2}{X^2} + \frac{\sigma_2^2}{Y^2} + \frac{\sigma_6^2}{S^2} = 1 \quad (23)$$

The above equation gives the Tsai-Hill criterion for failure for planar state of stress. From Tsai-Hill theory it is clear that it does not differentiate between tension and compression strengths for normal stresses. Infact, Tsai-Hill theory assumes same strengths in tension and compression. However, this situation does not occur in case of shear stresses. Thus, for normal stresses the theory represents a severe limitation that the sign of the normal stresses should be known a priori and the appropriate strength value should be used for normal stresses in the failure theory.

It should be noted that unlike maximum stress theory or maximum strain theory Tsai-Hill theory considers the interaction between three lamina strength parameters or interaction between stress components.

Further, it should be noted that Tsai-Hill theory is a unified theory and does not give the mode of failure like the maximum stress and maximum strain theory. However, one can make a guess of failure mode by calculating the quantities $\frac{\sigma_1^2}{X^2}, \frac{\sigma_2^2}{Y^2}$ and $\frac{\sigma_6^2}{S^2}$. The maximum of these three values can be said to give the mode of failure.

Note: The right hand side of Equation (20) or (23) is called as "failure index".

Permissions

All chapters in this book are published with permission under the Creative Commons Attribution Share Alike License or equivalent. Every chapter published in this book has been scrutinized by our experts. Their significance has been extensively debated. The topics covered herein carry significant information for a comprehensive understanding. They may even be implemented as practical applications or may be referred to as a beginning point for further studies.

We would like to thank the editorial team for lending their expertise to make the book truly unique. They have played a crucial role in the development of this book. Without their invaluable contributions this book wouldn't have been possible. They have made vital efforts to compile up to date information on the varied aspects of this subject to make this book a valuable addition to the collection of many professionals and students.

This book was conceptualized with the vision of imparting up-to-date and integrated information in this field. To ensure the same, a matchless editorial board was set up. Every individual on the board went through rigorous rounds of assessment to prove their worth. After which they invested a large part of their time researching and compiling the most relevant data for our readers.

The editorial board has been involved in producing this book since its inception. They have spent rigorous hours researching and exploring the diverse topics which have resulted in the successful publishing of this book. They have passed on their knowledge of decades through this book. To expedite this challenging task, the publisher supported the team at every step. A small team of assistant editors was also appointed to further simplify the editing procedure and attain best results for the readers.

Apart from the editorial board, the designing team has also invested a significant amount of their time in understanding the subject and creating the most relevant covers. They scrutinized every image to scout for the most suitable representation of the subject and create an appropriate cover for the book.

The publishing team has been an ardent support to the editorial, designing and production team. Their endless efforts to recruit the best for this project, has resulted in the accomplishment of this book. They are a veteran in the field of academics and their pool of knowledge is as vast as their experience in printing. Their expertise and guidance has proved useful at every step. Their uncompromising quality standards have made this book an exceptional effort. Their encouragement from time to time has been an inspiration for everyone.

The publisher and the editorial board hope that this book will prove to be a valuable piece of knowledge for students, practitioners and scholars across the globe.

Index

www.ingramcontent.com/pod-product-compliance
Lightning Source LLC
Chambersburg PA
CBHW082013190326
41458CB00010B/3175